Student Interactive Workbook

Human Biology

TENTH EDITION

Cecie Starr

Beverly McMillan

Prepared by

Jeff Taylor
SUNY Canton

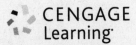
CENGAGE
Learning·

Australia • Brazil • Mexico • Singapore • United Kingdom • United States

CONTENTS

1   Learning About Human Biology 1

2   Chemistry of Life 12

3   Cells and How They Work 32

4   Tissues, Organs and Organ Systems 50

5   The Skeletal System 73

6   The Muscular System 90

7   Circulation: The Heart and Blood Vessels 103

8   Blood 118

9   Immunity and Disease 130

10   The Respiratory System 150

11   Digestion and Nutrition 165

12   The Urinary System 185

13   The Nervous System 196

14   Sensory Systems 217

15   The Endocrine System 231

16   Reproductive Systems 246

17   Development and Aging 265

18   Cell Reproduction 281

19   Observable Patterns of Inheritance 297

20   Chromosomes and Human Genetics 308

21   DNA, Genes, and Biotechnology 322

22   Genes and Disease: Cancer 345

23   Principles of Evolution 354

24   Principles of Ecology 367

25   Human Impacts on the Biosphere 379

Answer Key 392

PREFACE

Tell me and I will forget, show me and I might remember, involve me and I will understand.
—Chinese Proverb

The proverb outlines three levels of learning, each successively more effective than the method preceding it. The writer of the proverb understood that humans learn most efficiently when they involve themselves in the material to be learned. This study guide is like a tutor; when properly used it increases the efficiency of your study periods. The interactive exercises actively involve you in the most important terms and central ideas of your text. Specific tasks ask you to recall key concepts and terms and apply them to life; they test your understanding of the facts and indicate items to reexamine or clarify. Your performance on these tasks provides an estimate of your next test score based on specific material. Most important, though, this biology study guide and text together help you make informed decisions about matters that affect your own well-being and that of your environment. In the years to come, human survival on planet Earth will demand administrative and managerial decisions based on an informed biological background.

HOW TO USE THIS STUDENT WORKBOOK

Following this preface, you will find an outline that will show you how the study guide is organized and will help you use it efficiently. Each chapter begins with a title and a brief introduction to the topic, as well as figure references that proved excellent focal points for study. Interactive Exercises follow, wherein each chapter is divided into sections of one or more of the main headings within the textbook chapters. For easy reference to an answer or definition, each question and term in this study guide is accompanied by the appropriate text page(s). The Interactive Exercises begin with a list of Selected Words (other than boldfaced terms) chosen by the authors as those that are most likely to enhance understanding. In the text chapters, the selected words appear in italics, quotation marks, or roman type. This is followed by a list of Bold-faced, Page Referenced Terms that appear in the text. These terms are essential to understanding each workbook section of a particular chapter. Space is provided by each term for you to formulate a definition in your own words. Next is a series of different types of exercise that may include completion, short answer, true/false, fill-in-the-blanks, matching, choice, dichotomous choice, label and match, problems, labeling, sequencing, multiple choice, and completion of tables.

A Self-Quiz immediately follows the Interactive Exercises. This quiz is composed primarily of multiple-choice questions although sometimes we present another examination device or some combination of devices. Any wrong answers in the Self-Quiz indicate portions of the text you need to reexamine. A series of Chapter Objectives/Review Questions follows each Self-Quiz. These are tasks that you should be able to accomplish if you have understood the assigned reading in the text. Some objectives require you to compose a short answer or long essay while others may require a sketch or supplying correct words.

The final part of each chapter is named Integrating and Applying Key Concepts. It invites you to try your hand at applying major concepts to situations in which there is not necessarily a single pat answer and so none is provided in the chapter answer section. Your text generally will provide enough clues to get you started on an answer, but this part in intended to stimulate your thought and provoke group discussions.

A person's mind, once stretched by a new idea, can never return to its original dimension.
—Oliver Wendell Holmes

iv

STRUCTURE OF THIS STUDENT WORKBOOK

The following outline shows how each chapter in this student workbook is organized.

Chapter Number ——————▶

3

Chapter Title ——————▶

CELLS AND HOW THEY WORK

Introduction ——————▶

This includes a brief overview to the subject matter of the chapter.

Chapter 32 starts with a description of how cells are organized and connected to form organs, organ systems, and multicellular animals. The four types of animal tissues are described: epithelial, connective, muscle, and nerve. Vertebrate organ systems are introduced along with a more detailed description of how tissues are integrated into a specific organ system—the integumentary system.

Focal Points ——————▶

- Figure 32.2 [p.540] illustrates how cells are attached to each other to form coherent tissues.
- Figure 32.4 [p.541] shows various types of epithelium.
- Figure 32.5 and 32.6 [pp.542-543] illustrate connective tissues.
- Figure 32.8 [p.544] has images of muscle tissues.
- Figure 32.9 [p.545] shows a typical neuron.
- Figure 32.11 and 32.12 [pp.546-547] describe anatomical terms and outline the major human organ systems.
- Figure 32.13 [p.568] diagrams human skin structure.

Interactive Exercises ——————▶

The Interactive Exercises are divided into numbered sections by titles of main headings and page references. Each section begins with a list of author-selected words that appear in the chapter. This is followed by a list of important boldfaced, page-referenced terms from each section of the chapter. Each section ends with interactive exercises that vary in type and require constant interaction with the important chapter information

Self-Quiz ——————▶

This is a set of questions designed to provide a quick assessment of how well you understand the information from the chapter.

Chapter Objectives / ——————▶
Review Questions

This section provides a list of concepts that you need to master before proceeding to the next chapter. Page numbers from the text are provided should you be unable to answer the questions.

Integrating and ——————▶
Applying Key Concepts

These represent "big-picture" or "real-life" applications of the concepts presented in the chapter.

Answers to Self-Quiz ——————▶

Answers for all the Self-Quizzes can be found at the end of the student workbook. The answers are arranged by chapter number and main heading.

1

LEARNING ABOUT HUMAN BIOLOGY

INTRODUCTION

The theme of this chapter is the nature of life and its organization, as well as the methodology and way of thinking that are used to study life at all levels. The knowledge that you will acquire about the human body will allow you to critically analyze claims and advertisements you encounter concerning your health.

FOCAL POINTS

- All living things share basic characteristics which identify them as biological organisms [p.2]
- This chapter introduces the concept of homeostasis, [p.2, p.11] a common thread that will connect subsequent chapters.
- Evolution has brought about a diversity of life forms, including humans, each of which have a unique place in the natural world [p.3, Figure 1.3, *animated*]
- Nature is organized into many different levels, from the level of atoms to the level of the biosphere [pp.4-5, Figure 1.5, *animated*]
- Energy flows through organisms while nutrients cycle among them [p.5, Figure 1.6, *animated*]
- The Scientific Method of inquiry involves asking questions about observations in nature, creating a testable hypothesis, identifying variables and controls, collecting data or results, and drawing conclusions [p.7, Figure 1.8]
- A scientist thinks critically, basing conclusions on evidence, not opinion. Table 1.1 [p.8] is a guide to help you think critically.
- Scientific studies culminate in theories. Table 1.2 [p.9] lists four central theories of biological thought.
- The topic of infectious disease and public health is introduced [pp.10-11]

INTERACTIVE EXERCISES

Note: In the answer section of this book, certain molecules are often indicated by abbreviations. For example, deoxyribonucleic acid is DNA.

CHAPTER INTRODUCTION [p.1]

1.1. THE CHARACTERISTICS OF LIFE [p.2]

Boldfaced Terms

These terms are important; they are in boldface type in the chapter. Write a definition for each term in your own words without looking at the text. Next, compare your definition with that given in the chapter or in the text glossary. If your definition seems accurate, allow some time to pass and repeat this procedure until you can define each term rather quickly (how fast you answer is a gauge of your learning efficiency).

cell _____

homeostasis _____

_____

Choice

For examples 1–14, choose from the following characteristics of life: [p.2]

a. taking in and using energy and materials

b. sensing and responding to the environment

c. reproduction and growth

d. consist of one or more cell

e. maintaining homeostasis

1. _____ An animal must eat food in order to survive.

2. _____ DNA is passed to the next generation of a species.

3. _____ A driver stops when a traffic light turns red.

4. _____ The smallest units that are considered living.

5. _____ An organism changes as it ages in response to DNA instructions.]

6. _____ The internal environment of an organism stays within life-supporting ranges.

7. _____ A puppy grows up, mates and has puppies of her own.

8. _____ Units that contain DNA and use ATP.

9. _____ Human body temperature remains basically the same each day.

10. _____ A dog comes when its master whistles

11. _____ Means "staying the same"

12. _____ Every organism has a least one of these.

13. _____ Infancy, childhood, adolescence, adulthood

14. _____ Energy and molecules from protein in a meal are used to build muscle tissue.

1.2. OUR PLACE IN THE NATURAL WORLD [p.3]

Boldfaced Terms

primates _____

_____

vertebrates _____

_____

Hierarchical relationships

Fill in the blanks to represent the place of humans in the natural world. (Figure 1.3 in the text may be helpful.) Choose from the following words: [p.3]

humans   mammals   vertebrates   primates   animals

(broadest group) 1. _____ kingdom including humans and millions of other species

2. _____ animals with backbones

3. _____ vertebrates with body hair

4. _____ humans, apes and closely related mammals]

(narrowest group) 5. _____ primates with great manual dexterity and a brain providing verbal and analytical abilities

Listing

6. List the 4 kingdoms of life contained within the Domain Eukarya (see Figure 1.3). [p.3]

_____   _____   _____   _____

7. In a 5-kingdom system, all of the bacteria are placed in one kingdom. In a 3-domain system, they are split into 2 domains. List these 2 domains (see Figure 1.3).

_____   _____

1.3. LIFE'S ORGANIZATION [pp.4-5]

Boldfaced Terms

biosphere _____

_____

Matching

Choose the most appropriate description for each term. [pp.4-5]

1. _____ organ system
2. _____ cell
3. _____ community
4. _____ atoms and molecules
5. _____ ecosystem
6. _____ population
7. _____ tissue
8. _____ biosphere
9. _____ complex organism
10. _____ organ

A. Different tissues combined together to carry out a function in an organism

B. All parts of Earth's waters, crust, and atmosphere in which organisms live

C. The smallest living unit

D. Different organs working together in a coordinated manner

E. The populations of all species occupying the same area

F. Nonliving materials from which cells are built

G. A community and its surrounding environment

H. An individual composed of coordinated organ systems

I. A group of individuals of the same kind, such as all of Earth's humans

J. A group of cells organized to carry out a specific function, e.g., epithelium

Hierarchical Order

Arrange the following levels of organization in nature in the correct hierarchic order. Find the letter of the simplest level and write it next to the number 11; write the letter of the most complex level next to the number 20; and so forth. Refer to Figure 1.5. [pp.4-5]

11. _____
12. _____
13. _____
14. _____
15. _____
16. _____
17. _____
18. _____
19. _____
20. _____
21. _____

A. organ system
B. biosphere
C. organ
D. ecosystem
E. atom
F. complex organism
G. population
H. community
I. cell
J. tissue
K. molecule

Fill-in-the-Blanks [pp.4-5]

Energy enters the biosphere from the (22) _____ Plants and other organisms that use this energy to make

food molecules by photosynthesis are called (23) _____. Animals that eat plants or other animals to obtain

energy and nutrient molecules are called (24) _____. Bacteria and fungi obtain the materials and energy they

need by (25) _____ the remains of other organisms, thus recycling substances back to the

photosynthesizers. Because of these interconnections, ecosystems can be thought of as interconnected

(26)_____ of life in which events in one part impact the entire ecosystem.

1.4. USING SCIENCE TO LEARN ABOUT THE NATURAL WORLD [pp.6-7]

1.5. CRITICAL THINKING IN SCIENCE AND LIFE [p.8]

1.6. SCIENCE IN PERSPECTIVE [p.9]

Boldfaced Terms

scientific method _____

_____

hypothesis _____

_____

experiment _____

_____

variable _____

_____

control group _____

_____

sampling error_____

_____

critical thinking _____

_____

fact _____

_____

opinion _____

_____

scientific theory _____

_____

Complete the Table

1. Complete the following table of concepts important to understanding the scientific method of problem solving. Choose from *experiment, variable, observation, prediction, control group,* and *hypothesis.* [pp.6-7]

Concept	Definition
a.	This, along with curiosity, is the source of scientific investigations.
b.	A tentative explanation for a phenomenon that can be tested objectively
c.	A statement of what one should be able to observe about a problem if a hypothesis is valid
d.	A procedure designed to test a prediction stemming from a hypothesis
e.	The control group is identical to the experimental group except for this key factor
f.	Used in scientific experiments as a standard to which the experimental group is compared

Sequence

Arrange the following steps of the scientific method in the correct sequence. Find the letter of the first process and write it next to the number 2, and so forth, finishing with the letter of the final process next to the number 7. [p.6]

2. ____

3. ____

4. ____

5. ____

6. ____

7. ____

A. develop a hypothesis

B. test the accuracy of prediction

C. repeat tests or devise new tests of the hypothesis

D. make a prediction

E. objectively analyze and report the test results and conclusions

F. observe some aspect of the natural world and identify a question to be explored

Short Answer

8. Distinguish between *fact* and *opinion*. [pp.8-9] _____

_____

_____

9. List four ways to think critically. [pp.8-9] _____

_____

_____

10. Distinguish between hypothesis and theory. [pp.8-9] _____

_____

_____

True/False

In the blank beside each statement below, write a "T" (true) or an "F" (false). [p.9]

11. _____   A theory in science is simply a person's idea about something.

12. _____   Once a theory is accepted by scientists, its accuracy is never challenged again.

13. _____   A theory explains a broad range of related events and observations regarding the natural world.

14. _____   With time and better technology, science will be able to investigate and answer many questions that humans ask.

15. _____   Science requires a strictly objective mindset.

16. _____   Science does not involve value judgments.

17. _____   The scientific community should take complete responsibility for guiding applications of scientific knowledge.

Dichotomous Choice

Circle the correct choice within each set of brackets. [p.8]

A study shows that, as a group, women who jog regularly develop breast cancer less often than women who never jog. This data reveals that jogging (18) (causes / is correlated to) a lower incidence of breast cancer. The manufacturer of a home jogging machine publishes the results of the study on breast cancer in order to encourage women to buy its product. Checking the accuracy of the commercial and questioning the motivation are important aspects of (19) (opinionated / critical) thinking. The manufacturer also claims that using the jogging machine will result in a happier lifestyle. Because this information cannot be verified, it is an (20) (fact / opinion).

1.7. LIVING IN A WORLD OF INFECTIOUS DISEASE [pp.10-11]

1.8. HOMEOSTASIS [p.11]

Boldfaced Terms

emerging diseases _____

_____

antibiotic _____

_____

Matching

Select the description, that best matches the term. [pp.10-11]

1. ____ Pathogen
2. ____ Ebola virus
3. ____ SARS virus
4. ____ Antibiotic
5. ____ Lyme disease
6. ____ Antiviral drugs
7. ____ Antibiotic resistance
8. ____ West Nile virus
9. ____ African sleeping sickness
10. ____ Emerging disease

A. a cause of encephalitis

B. causes severe respiratory disease

C. interfere with the viral life cycle

D. due to overexposure, many bacteria no longer respond to some antibiotics

E. causes "hemorrhagic fever" and massive bleeding

F. organism that can cause disease

G. caused by *Trypanasoma brucei*, a microscopic protozoan

H. caused by a pathogen that was formerly not infectious or was present only in a limited area

I. many of these substances are produced by bacteria and fungi

J. a bacterial disease transmitted by a tick bite

True or False

Write *True* or *False* in the space next to each question. [pp.10-11]

11. _____ Doctors were able to treat patients with antibiotics during the Spanish Flu pandemic of 1918.

12. _____ Antibiotics are effective treatments for viral infections such as those caused by Ebola and West Nile viruses.

13. _____ The addition of antibiotics to soaps, hand creams and other everyday products has helped to create resistant strains of bacteria.

14. _____ West Nile Encephalitis is considered an emerging disease because it was little known and not widespread until the last few years.

SELF-TEST

___ 1. Which of these is NOT a characteristic unique to humans? [p.3]

    a. use of a sophisticated verbal language
    b. development of advanced technology
    c. use of tools
    d. wide variety of social behavior

___ 2. About 12 to 24 hours after a meal, a person's blood sugar level normally varies from about 60 to 90 mg per 100 ml of blood, although it may attain 130 mg/100 ml after meals high in carbohydrates. The maintenance of blood sugar level within a fairly narrow range despite uneven intake of sugar is due to the body's ability to maintain a state of _____. [p.2]

    a. adaptation
    b. diversity
    c. metabolism
    d. homeostasis

___ 3. _____ is a change in the body plan and functioning of organisms through successive generations. [p.3]

    a. Homeostasis
    b. Reproduction
    c. Metabolism
    d. Evolution

___ 4. Webs of life including producers, consumers, and decomposers that function in the physical environment are called _____. [p.4-5]

    a. populations
    b. biomes
    c. ecosystems
    d. food chains

___ 5. The information-gathering system used by scientists to gain knowledge about the natural world is _____. [p.6-7]

    a. the scientific method
    b. inductive reasoning
    c. deductive reasoning
    d. the "if–then" process

___ 6. The experimental group and the control group are identical except for _____. [p.7]

    a. the results they give in an experiment
    b. the variable being studied
    c. the number of test subjects
    d. the number of experiments performed on each group

___ 7. Which of these is true of a hypothesis? [pp.6, 9]

    a. It is the same as a theory.
    b. It is the same as a prediction.
    c. It is an educated guess used to launch an investigation.
    d. It must always be proven true.

___ 8. Which of the following is NOT an example of critical thinking? [p.8]

    a. gathering information from reliable sources
    b. relying on the opinions of others
    c. using facts that can be independently verified
    d. being open to changing your point of view

___ 9. Which of the following is a Domain? [p. 3]

    a. Bacteria
    b. Protists
    c. Fungi
    d. Animals

___ 10. Which of the following lists a correct order of the organization of life from simplest to more complex? [p.4]

    a. tissue, cell, organ, organ system, molecule
    b. molecule, cell, tissue, organ system, organ
    c. molecule, tissue, organ system, organ, cell
    d. molecule, cell, tissue, organ, organ system

____ 11. Which of the following lists a correct order of the organization of life from simplest to more complex? [pp.4-5]

    a. organism, community, biosphere, population, ecosystem
    b. community, ecosystem, population, organism, biosphere
    c. organism, population, community, ecosystem, biosphere
    d. population, organism, community, ecosystem, biosphere

____ 12. Disease may be caused by _____. [p.10]

    a. bacteria
    b. viruses
    c. parasites
    d. all of the above

____ 13. Which of the following is INCORRECT? [p.11]

    a. Antibiotics are made by bacteria and fungi.
    b. Antibiotics work by interfering with cell processes.
    c. Antibiotics are effective against viral infections
    d. Antibiotics can trigger allergic responses

CHAPTER OBJECTIVES/REVIEW QUESTIONS

This section lists general and detailed chapter objectives that can be used as review questions. You can make maximum use of these items by writing answers on a separate sheet of paper. Fill in answers where blanks are provided. To check for accuracy, compare your answers with information given in the chapter or glossary.

1. A special molecule called _____ directs the growth and development of organisms. [p.2]

2. In addition to growth and development, what other characteristics do all living things share? [p.2]

3. To sustain life, body systems must maintain a dynamic state of internal equilibrium called _____. [p.2]

4. A(n) _____ is an organized living unit that can survive and reproduce itself, using DNA instructions and the necessary energy and materials. [p.2]

5. List the three domains of living things. Which two include bacteria, the simplest organisms? To which domain do humans belong? [p.3]

6. Humans, goldfish, lizards and apes are placed in a group of animals called vertebrates. Of these, only apes and humans are in the small subdivision of vertebrates called primates. What does this grouping suggest about the evolutionary relatedness of humans to these other animals? [p.3]

7. Explain how the process of evolution has created both similarity and diversity of life forms. [p.3]

8. Name and arrange the levels of organization in nature, from the least inclusive to the most inclusive (nonliving cellular components to biosphere). [pp.4-5]

9. Explain how the actions of producers, consumers, and decomposers create interdependency among organisms. [pp.4-5]

10. List and explain the seven steps that describe what scientists generally do when they proceed with an investigation. [p.6]

11. Why is a conclusion based on data gathered from a small group of experimental subjects less reliable than one supported by data from a large group? [p.7]

12. What is the function of the control group in an experiment? How does the control group differ from the experimental group? What is a variable? [p.7]

13. Define "critical thinking" and list ways in which a piece of information is evaluated critically. [p.8]

14. How does a scientific theory differ from a scientific hypothesis? [p.9]

15. What attribute of science leads to controversy over the potential use of scientific findings? [p.9]

16. Explain why no theory is an "absolute truth". [p.9]

17. What is a pathogen? What types of organisms are considered pathogens? [p.10]

18. How do antibiotic and antiviral drugs differ in function? [p.10]

19. Define antibiotic resistance and list possible causes of this phenomenon. [p.11]

INTEGRATING AND APPLYING KEY CONCEPTS

1. Consider the importance of the sun to the energy balance of the biosphere. How would prolonged lack of sunshine affect the organisms on earth? Be sure to discuss the implications on producers, consumers and decomposers.

2. What sorts of topics do scientists usually regard as not testable by the methods they generally use?

3. What might the consequences be if a particular bacterial pathogen becomes resistant to all known antibiotics?

2

CHEMISTRY OF LIFE

INTRODUCTION

Our bodies are made up of a dozen main elements with traces of a few others. All of the functions of the human body revolve around the interactions of these chemicals. This chapter looks at the important chemical elements, molecules and compounds in the human body and helps to simplify the understanding of how we function at the chemical level.

FOCAL POINTS

- Table 2.1 [p.19] illustrates several different ways to represent the same molecule. It is important to understand these different representations as one or more of them will appear throughout the book referring to different molecular compounds.
- Figure 2.12, [p.24] illustrates the pH scale using common, everyday items. Maintaining a stable pH is critical to the survival of cells and tissues. Familiarity with the pH scale will be of value in subsequent chapters.
- Figures 2.17 [p.28], 2.20, [p.30], 2.25, animated [p.33] and 2.30, [p.36] show chemical structures of the four main types of biological molecules – carbohydrates, lipids, proteins, and nucleic acids. It is important to have a general mental image of these types of molecules in order to understand their structures and functions.

INTERACTIVE EXERCISES

CHAPTER INTRODUCTION [p. 15]

2.1. ATOMS AND ELEMENTS [pp.16-17]

2.2. PET SCANNING - USING RADIOISOTOPES IN MEDICINE [p.17]

2.3. CHEMICAL BONDS: HOW ATOMS INTERACT [p.18-19]

Boldfaced Terms

element_____

_____

atom_____

_____

isotope _____

_____

radioisotope _____

_____

tracer _____

_____

chemical bond _____

_____

molecule _____

_____

compound _____

_____

mixture _____

_____

Matching

Match each of the following chemistry terms with its correct definition. [pp.16-19]

1. _____ atom
2. _____ atomic number
3. _____ electrons
4. _____ element
5. _____ isotopes
6. _____ mass number
7. _____ neutrons
8. _____ tracer
9. _____ trace element
10. _____ chemical bond
11. _____ inert
12. _____ nucleus
13. _____ compound
14. _____ radioactivity
15. _____ protons
16. _____ radioisotope

A. The smallest unit that has the properties of a given element

B. The number of protons in the nucleus of one atom of an element; defining characteristic of an element

C. Subatomic particles in the nucleus of an atom that have no charge

D. An element important to the body that represents less than 0.01 percent of body weight

E. Energy emitted from unstable isotopes

F. Positively charged subatomic particles in the nucleus of an atom

G. Area within an atom where protons and neutrons are found

H. Unstable isotope that stabilizes itself by spontaneously emitting energy and particles

I. Formed by two or more elements combined in fixed proportions

J. Negatively charged subatomic particles that move rapidly around the nucleus of an atom

K. Fundamental form of matter; cannot be broken down into other substances by ordinary processes

L. Forms of an element, the atoms of which contain a different number of neutrons than other forms of the same element

M. A substance with a radioisotope attached so that its movement can be tracked

N. Combined number of protons and neutrons in the nucleus of an atom

O. A union between the electron structures of atoms

P. An atom having no vacancies in its outer shell

Dichotomous Choice

Circle the correct choice within each set of parentheses. [pp. 16-19]

17. To determine the number of neutrons in an atom, you must know the atomic number and the (number of electrons / mass number).

18. For an isotope of carbon, the number that would change would be the (atomic number / mass number).

19. PET Scanning is a useful tool in medicine because it utilizes the fact that normal cells are (less/more) active than cancer cells.

20. An atom is most stable when its (inner/outer) shell is filled.

Short Answer

21. Oxygen gas consists of two atoms of oxygen bonded together. Water consists of two atoms of hydrogen and one atom of oxygen bonded together. Which are molecules? Which are compounds? Explain. [p.19]

_____

_____

_____

_____

Elimination

For each statement below, cross out the choice that does NOT fit. [pp. 18-19]

22. For most atoms, eight electrons can fit in the outer (shell / orbital / energy level).

23. Chemical bonding occurs between the components of a (molecule / mixture / compound).

24. The most abundant elements in the human body include (oxygen / hydrogen / calcium / nitrogen).

Multiple Choice

The following statements pertain to the two chemical reactions symbolized below. Circle the choice(s) that correctly completes each sentence. (More than one choice may be circled for each statement.) [p. 19]

$$2Na + Cl_2 \longrightarrow 2NaCl \qquad 2H_2 + O_2 \longrightarrow 2H_2O$$

25. The two reactions above are (chemical equations / physical reactions / molecular processes).

26. Na and H_2 are both (products / molecules / reactants).

27. NaCl and H_2O are both (products / reactants / elements).

28. The equation on the right must include two H_2 and two H_2O in order to be (reversible / properly bonded / balanced).

2.4. IMPORTANT BONDS IN BIOLOGICAL MOLECULES [pp.20–21]

2.5. WATER: NECESSARY FOR LIFE [pp.22-23]

Boldfaced Terms

ion _____

_____

ionic bond _____

_____

covalent bond _____

_____

hydrogen bond _____

_____

biological molecule _____

_____

hydrophilic _____

_____

hydrophobic _____

_____

solvent _____

_____

solute _____

_____

Short Answer

1. How do hydrogen bonds hold atoms together? [p.21] _____

    _____

    _____

    _____

2. Why does hydrogen gas have a single covalent bond, oxygen gas have a double covalent bond and nitrogen gas have a triple covalent bond? [p.20] _____

    _____

    _____

    _____

Identification

3. The figure below depicts an ionic bond between a sodium atom and a chlorine atom forming sodium chloride (NaCl). Using this model, draw the molecule representing calcium chloride given that calcium's atomic number is 20. [p.20]

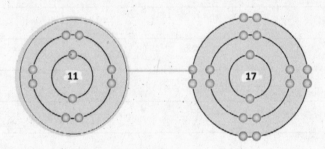

4. Using the figure in question 3, draw the transfer of electron(s) (by arrows) that takes place when positive aluminum (Al^{3+}) and negative chloride (Cl^-) ions form ionic bonds to create a molecule of $AlCl_3$ (aluminum chloride). [p.20]

Choice

The following statements pertain to the two chemical reactions symbolized below. Circle the choice(s) that correctly completes each sentence. (More than one choice may be circled for each statement.) [pp. 20-21]

$$2Na + Cl_2 \longrightarrow 2NaCl \qquad 2H_2 + O_2 \longrightarrow 2H_2O$$

5. The reaction on the left illustrates ionic bonding. In this type of reaction, one or more electrons are (shared / donated / accepted).

6. The component atoms of NaCl stay together because they have (the same / opposite / no) charge.

7. The reaction on the right illustrates bonding in which electrons are shared, although the charge of the electrons is not distributed equally. The bond that results is (covalent / polar / nonpolar).

8. If H_2O were drawn as a structural formula, a line would illustrate each (covalent / ionic / hydrogen) bond.

9. The two H_2Os in the second reaction could be attracted to each other by weak (covalent / ionic / hydrogen) bonds.

Dichotomous Choice

Circle one of two possible answers given between parentheses in each statement. [pp. 22-23]

10. Your (body / sweat) is ~99% water.

11. The polarity of water molecules allows them to form (hydrogen/covalent) bonds with one another and with other polar substances.

12. Polar molecules are attracted to water and hence are said to be (hydrophobic/hydrophilic).

13. The polarity of water repels oil and other nonpolar substances, which are (hydrophobic/hydrophilic).

14. Hydrogen bonds between water molecules enable water to absorb a great deal of (cold/heat) energy before it significantly warms or evaporates.

15. When the amount of heat energy is raised sufficiently, hydrogen bonds between water molecules break apart and (re-form/do not re-form); water then evaporates.

16. Sweating followed by evaporation results in the body (gaining/losing) heat energy.

17. Water is a superb (solute/solvent).

18. Substances dissolved in water are (solutes/solvents).

19. A substance is said to (precipitate/dissolve) as clusters of water molecules form around its individual ions or molecules.

2.6. HOW ANTIOXIDANTS PROTECT CELLS [p.23]

2.7. ACIDS, BASES AND BUFFERS: BODY FLUIDS IN FLUX [pp.24-25]

Boldfaced Terms

free radical _____

_____

antioxidant _____

_____

hydrogen ion _____

_____

hydroxide ion _____

_____

pH scale _____

_____

acid _____

_____

base _____

_____

buffer _____

_____

salts _____

_____

Short Answer

1. Explain how free radicals are produced and why they are a potential threat to human health. [p.23]

_____

_____

_____

_____

2. What is an antioxidant? Name some antioxidants available in our diet. [p.23]

_____

_____

_____

_____

_____

Labeling and Analysis

```
  |    |    |    |    |    |    |    |    |    |    |    |    |    |    |
  0    1    2    3    4    5    6    7    8    9   10   11   12   13   14
```

(3)_____   (4)_____

3-4. In the blanks provided above, label the basic (alkaline) and acidic ends of the pH scale above. [p.24]

5. Circle the number on the scale above at which the concentrations of H^+ and OH^- are equal (neutrality). [p.24]

6. A solution with a pH of 8 is ten times more basic (alkaline) than one with a pH of 7. A solution with a pH of 9 is _____ more basic than one with a pH of 7. [p.24]

7. Gastric fluid with a pH of 2 is _____ more acidic than pure water with a pH of 7. [p.24]

8. Blood has a pH of 7.3-7.5. It is slightly _____. [p.24]

$$H_2CO_3 \longrightarrow HCO_3^- + H^+ \qquad\qquad HCO_3^- + H^+ \longrightarrow H_2CO_3$$

9. The equations above represent the reversible reactions of a _____ system. [p.25]

10. _____ is the reactant in the left equation and the product on the right. _____ + H^+ are the products on the left and the reactants on the right. [p.25]

11. The equation on the (left / right) occurs when the pH of blood and tissue fluid begins to become more acidic. This occurs in order to (add / remove) H^+. [p.25]

12. The equation on the (left / right) occurs when the pH of blood and tissue fluid begins to become more basic. This occurs in order to (add / remove) H^+. [p.25]

13. This type of system helps to prevent uncontrolled shifts in pH such as a severe decrease in blood pH called (acidosis / alkalosis), or a severe increase in blood pH called (acidosis / alkalosis). [p.25]

14. If blood pH drops to 7.0, (no change / coma) will occur. If it rises above 7.5, (no change/ death) may occur. [p.25]

$$H_2CO_3 \qquad\qquad HCO_3^- \qquad\qquad H^+$$

(15)_____   (16)_____   (17)_____

15-17. In the blanks provided above, label the buffer, acid and hydrogen ion components of the reaction. [p.25]

18. A(n) (acid / base / salt) is an ionic compound that releases ions other than H^+ and OH^-. [p.25]

19. A buffered fluid's pH (decreases/stays constant/increases) even when a base is added. [p.25]

Short Answer

20. Explain how milk of magnesia acts as an antacid in the stomach. [pp.24-25] _____

_____

_____

_____

21. What is a buffer system's action on hydrogen ions? [p.25] _____

_____

_____

_____

22. What forms when a strong base and a strong acid interact? [p.25] _____

_____

_____

_____

23. Explain why carbon dioxide is an important part of the body's buffer system. [p.25] _____

_____

_____

_____

Complete the Table

24. Complete the table with the missing information. [p. 24]

Substance	pH Value	Acid or Base
rainwater	a.	b.
seawater	c.	d.
acid rain	e.	f.
gastric fluid	g.	h.
Tums	i.	j.

2.8. MOLECULES OF LIFE [pp.26–27]

Boldfaced Terms

organic compound _____

_____

functional group _____

_____

enzymes _____

_____

condensation reaction _____

20   Chapter Two

polymer _____

_____

monomers _____

_____

hydrolysis _____

_____

True/False

If a statement is true, write a "T" in the blank. If it is false, underline the incorrect word and write the correct word in the blank. [pp. 26-27]

1. _____ Organic compounds have hydrogen and often other elements bonded to atoms of carbon by covalent bonds.

2. _____ The four classes of biological molecules are carbohydrates, lipids, proteins, and water.

3. _____ Carbon forms two covalent bonds (2 pairs of shared electrons).

4. _____ Organic molecules can be chains or rings.

5. _____ A carbon backbone with only hydrogen atoms attached is a carbohydrate.

6. _____ Atoms or clusters of atoms that are covalently bonded to the carbon backbone and influence the chemical behavior of an organic compound are called enzymes.

7. _____ Estrogen and testosterone have the same functional groups located in different places on the molecules.

8. _____ Rearrangement of an organic molecule constitutes the giving up of a functional group by one molecule and the immediate acceptance of it by another molecule.

Short Answer

9. State the general role of enzymes as they relate to reactions. [p.27] _____

_____

_____

_____

10. What is the byproduct of a condensation reaction and what does this byproduct have to do with a hydrolysis reaction? Explain briefly what takes place in both reactions. [p.27] _____

_____

_____

_____

11. By what process would the monomers in a polymer be separated? [p.27] a. _____
    Is an enzyme required for this process? [p.27] b. _____

2.9. CARBOHYDRATES: PLENTIFUL AND VARIED [pp.28–29]

Boldfaced Terms

carbohydrates _____

_____

monosaccharide _____

_____

oligosaccharide _____

_____

polysaccharides _____

_____

Complete the Table [pp.28-29]

1. In the following table, enter the name of the carbohydrate described by its carbohydrate class and function(s). Select from *glucose, sucrose, lactose, glycogen, cellulose, starch,* and *deoxyribose.*

Carbohydrate	Carbohydrate Class	Function
a.	Oligosaccharide	Most plentiful sugar in nature; table sugar
b.	Monosaccharide	Main energy source for body cells; building block of many organic compounds
c.	Polysaccharide	Tough, insoluble structural material in plant cell walls; fiber in our diet
d.	Monosaccharide	Five-carbon sugar; occurs in DNA
e.	Oligosaccharide	Sugar present in milk
f.	Polysaccharide	Storage form of sugar in animals, including humans
g.	Polysaccharide	Storage form of glucose in plants

2.10. LIPIDS: FATS AND THEIR CHEMICAL RELATIVES [pp.30–31]

Boldfaced Terms

lipid _____

_____

fat _____

_____

fatty acid _____

_____

triglyceride _____

_____

phospholipid _____

_____

sterol _____

_____

Labeling

1. In the appropriate blanks, label the molecule shown as either *saturated, unsaturated* or *polyunsaturated.* [p.30]

a. _____

b. _____

c. _____

a
stearic acid

b
oleic acid

c
linolenic acid

Short Answer

2. Name some important molecules in our bodies that are made from cholesterol. [p.31]

_____

_____

3. Describe the structure of the cell membrane. [p.31]

_____

_____

Matching

Choose the most appropriate description for each term. [pp. 30-31]

4. _____ fatty acids

5. _____ triglycerides

6. _____ phospholipids

7. _____ trans fatty acids

8. _____ sterols

A. Partially saturated (hydrogenated)

B. Up to 36 carbons long with unsaturated or saturated tails

C. Cholesterol, steroid hormones

D. Main components of cell membranes

E. Body's richest source of energy; e.g., butter, lard, and oils

2.11. PROTEINS: BIOLOGICAL MOLECULES WITH MANY ROLES [pp.32–33]

2.12. A PROTEIN'S SHAPE AND FUNCTION [pp.34–35]

Boldfaced Terms

protein _____

_____

amino acid _____

_____

peptide bond _____

_____

polypeptide chain _____

_____

primary structure _____

_____

_____

lipoprotein _____

_____

glycoprotein _____

_____

Labeling

For questions 1–3, write the name of the major parts of every amino acid. Choose from *R group*, *carboxyl group* and *amino group*. [p.32]

1. _____

2. _____

3. _____

valine (val)   tryptophan (trp)   methionine (met)

4. _____   Which of these three parts differs from one amino acid to another, and determines the chemical and physical properties of the amino acid?

5. _____   Which type of bond connects the three parts to the carbon atom in the middle?

Matching

Choose the most appropriate description for each term. [pp. 33-34]

6. ____ primary protein structure

7. ____ secondary protein structure

8. ____ tertiary protein structure

9. ____ quaternary protein structure

A. Unique sequence of amino acids in the polypeptide chain of a specific protein

B. Coils, sheets and loops fold even further, some folding into a hollow "barrel"

C. Polypeptide chain twists, bends, loops and folds

D. Bending and looping of the polypeptide chain, caused by the interaction of R group

10. Glycoproteins and lipoproteins are both modified proteins. How are they different from each other? [pp.34,35]

_____

_____

_____

11. Describe protein denaturation. [p.35] _____

_____

_____

_____

2.13. NUCLEOTIDES AND NUCLEIC ACIDS [p.36]

2.14. FOOD PRODUCTION AND A CHEMICAL ARMS RACE [p.37]

Boldfaced Terms

nucleotide _____

_____

ATP _____

_____

coenzyme _____

_____

nucleic acid _____

_____

DNA _____

_____

RNA _____

_____

Labeling

For questions 1–3, write the name of the major parts of a nucleotide. Choose from *sugar*, *phosphate group* and *base*. [p.36]

1. _____

2. _____

3. _____

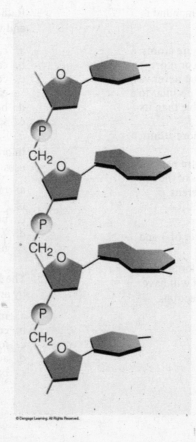

Short Answer

4. What is the difference between DNA and RNA? [p.36] _____

_____

_____

_____

Identification

5. In the accompanying diagram of a single-stranded nucleic acid molecule, encircle as many complete nucleotides as possible. How many complete nucleotides are present? [p.36]

Matching

Choose the most appropriate description for each term. [p. 36]

6. ____ adenosine triphosphate (ATP)

7. ____ RNA

8. ____ DNA

A. Single nucleotide strand; plays a key role in processes by which genetic instructions are used to build the body's proteins

B. Nucleotide that provides energy for cellular reactions by the transfer of a phosphate group

C. Helical double-nucleotide strand; encodes genetic information with base sequences

Short Answer

9. What is the benefit of chemical agents used to protect crops? What are the drawbacks? [p.37]

_____

_____

_____

SELF-QUIZ

___ 1. A molecule is _____; a compound is _____. [p.19]

  a. a combination of two or more atoms; a molecule consisting of two or more elements in proportions that never vary
  b. the smallest unit of matter peculiar to a particular element; less stable than its constituent atoms
  c. a combination of two or more atoms; a very large molecule
  d. a carrier of one or more extra neutrons; a substance in which atoms of the same element are present in different proportions

___ 2. If lithium has an atomic number of 3 and an atomic mass of 7, it has _____ neutrons in its nucleus; if a chlorine atom has an atomic number of 17, it will have _____ shells containing electrons. [pp.16,18]

  a. three; two
  b. four; two
  c. three; three
  d. four; three
  e. seven; three

___ 3. Radioactive decay occurs spontaneously and will [p.17]

  a. cause an atom to increase in size.
  b. cause an atom to decrease in size.
  c. convert an atom into a different element.
  d. bond two atoms together.
  e. split two atoms apart.

___ 4. In order to detect cancer, a PET scanner uses _____. [p.17]

  a. radioisotopes
  b. _____ glucose injections
  c. buffers
  d. X-rays
  e. nucleotides

___ 5. The loss or gain of one or more electrons by an atom creates a(n) _____. [p.20]

  a. molecule
  b. compound
  c. ion
  d. acid
  e. base

6. Polar molecules attracted to water are said to be _____. [p.22]
   a. hydrophobic
   b. hydrophilic
   c. cohesive
   d. free radicals
   e. acidic

7. Amino, carboxyl, phosphate, and hydroxyl are examples of _____; _____ form the structural elements of bones and muscles. [pp.26,32]
   a. functional groups; carbohydrates
   b. sugar units; nucleic acids
   c. functional groups; proteins
   d. coenzymes; lipids

8. Fatty acids with two or more double bonds are classified as _____. [p.30]
   a. unsaturated
   b. saturated
   c. polyunsaturated
   d. glycosidic

9. Hydrolysis could be correctly described as the _____; glycogen, plant starch, and cellulose are _____. [pp.27,29]
   a. heating of a compound in order to drive off its excess water and concentrate its volume; proteins
   b. breaking of a polymer into its subunits by adding water molecule components to separated monomers; carbohydrates
   c. linking of two or more molecules by the removal of one or more water molecules; lipids
   d. constant removal of hydrogen atoms from the surface of a carbohydrate; nucleic acids

10. Genetic information is encoded in the sequence of the bases of _____; molecules of _____ function in processes using genetic instructions to build the body's proteins. [p.36]
    a. DNA; DNA
    b. DNA; RNA
    c. RNA; DNA
    d. RNA; RNA

CHAPTER OBJECTIVES/REVIEW QUESTIONS

This section lists general and detailed chapter objectives that can be used as review questions. You can make maximum use of these items by writing answers on a separate sheet of paper. Fill in answers where blanks are provided. To check for accuracy, compare your answers with information in the chapter or glossary.

1. A(n) _____ is a fundamental unit of all matter. [p.16]

2. The body's ability to manage changes that disturb its chemistry helps to maintain the internal stability called _____. [p.15]

3. How does atomic number differ from mass number? [p.16]

4. ^{12}C and ^{14}C represent _____ of carbon. [p.17]

5. _____ are unstable and tend to spontaneously emit subatomic particles or energy in order to achieve stability. [p.17]

6. _____ (PET) utilizes radioactive tracers to help make diagnoses and observe brain activity. [p.17]

7. The maximum number of electrons that a shell around the nucleus of an atom beyond helium may contain is _____. [p.18]

8. Atoms with an unfilled outer shell tend to form chemical _____ with other atoms and so fill their outer shell. [p.19]

9. Atoms, such as helium, argon and neon have no vacancies in their shells and are considered _____. [p.19]

10. When chemical bonding joins atoms, the new structure is a _____. [p.19]

11. A(n) _____ bond is an attraction between two oppositely charged ions. [p.21]

12. "A pair of electrons shared between two atoms" defines a(n) _____ bond. [p.20]

13. The bond between two atoms that do not share electrons equally is a _____. [p.20]

14. Describe how hydrogen bonds are formed; cite one example of a large molecule in which many weak hydrogen bonds firmly hold its two sides together. [p.21]

15. Polar molecules are attracted to water and are said to be _____. [p.22]

16. Nonpolar molecules are repelled by water and are known as _____. [p.22]

17. Describe how an antioxidant works. [p.23]

18. A substance that releases H^+ when it dissolves in water is a(n) _____; any substance that accepts H^+ when it dissolves in water is a(n) _____. [p.24]

19. Describe the structure and use of the pH scale. [p.24]

20. Define *buffer system;* describe how the carbonic acid/bicarbonate buffer system works in the blood. [p.25]

21. Describe what happens to the blood in a case of acidosis. [p.25]

22. Acids often combine with bases to form _____ and water. [p.25]

23. List the four main types of biological molecules that represent the molecules characteristic of life. [p.26]

24. A(n) _____ is a carbon backbone with only hydrogen atoms attached. [p.26]

25. –OH, –COOH, and –NH$_3$ are examples of _____ groups. [p.26]

26. Define the terms *condensation* and *hydrolysis,* and *polymer* and *monomer,* and explain the first two in terms of their relationship with the second two. [p.27]

27. Cells use carbohydrates to _____ [p.28]

28. List the three classes of carbohydrates; how is each distinguished from the other two? [pp.28,29]

29. What is the structural difference between saturated, unsaturated and polyunsaturated fatty acids? [p.30]

30. The main materials of cell membranes are _____. [p.31]

31. Lipids that have no fatty acid tails are _____. [p.31]

32. List some positive functions of cholesterol. [p.31]

33. The bond that connects amino acids to form proteins is a(n) _____ bond. [p.33]

34. List the major functions of proteins; all proteins are constructed from about 20 different kinds of _____ acids. [p.32]

35. One group of proteins, the _____, make metabolic reactions proceed much faster than they otherwise would. [p.32]

36. Briefly describe the four levels of a protein's structure. [pp.33-34]

37. The most common protein in the body is _____. [p.34]

38. The loss of a protein's three-dimensional shape following disruption of bonds is called _____. [p.35]

39. How are nucleotides related to nucleic acids? Cite the names and functions of two major nucleic acids important to life. [p.36]

40. List some of the negative affects that pesticides have on humans. [p.37]

INTEGRATING AND APPLYING KEY CONCEPTS

1. Explain what would happen if water were a nonpolar molecule instead of a polar molecule. Would water be a good solvent for the same kinds of substances? Would the nonpolar molecule's heat capacity likely be higher or lower than that of water? Would its ability to form hydrogen bonds change?

2. Whereas plants store their energy as starch, the long-term energy storage molecule of animals is fat. While starch attracts some water, fat is hydrophobic and it thus stores more energy in a smaller space. Why do you think that fat is a better energy storage molecule for animals in spite of the fact that it is more difficult to break down than is starch?

3. Most DNA information tells the cell how to make proteins, while none of this information represents instructions for other types of organic molecules. What type of protein allows reactions to occur that make or break down other organic molecules? Explain why damage to DNA instructions for making a specific enzyme may result in a genetic disease.

3

CELLS AND HOW THEY WORK

INTRODUCTION

In chapter 1, the cell was defined as the fundamental unit of life. This chapter examines the two basic cell types, prokaryote and eukaryote, and the specialized organelles found in eukaryotic cells that carry out life functions. The structure of the plasma membrane, and how it controls what enters and leaves the cell, is explained. Cellular metabolism and energy usage and production are also considered in this chapter. Understanding the interrelated structure and function of cellular organelles will provide you with the foundation needed later on in the textbook.

FOCAL POINTS

- Figure 3.4 animated [p.43] illustrates the basic phospholipid structure of the cell membrane.
- Figure 3.5 animated [p.44] demonstrates the various organelles found in an animal cell.
- Figure 3.6 animated [p.45] compares light and electron microscopes and the images produced by them.
- Figure 3.7 animated [p.46] visually demonstrates the components of the plasma membrane.
- Figure 3.8 animated [p.47] demonstrates selective permeability of the plasma membrane.
- Figure 3.11 animated [p.49] shows how the structure of the nuclear envelope is related to its function.
- Figure 3.12 animated [pp.50-51] illustrates the interrelationships between components of the endomembrane system.
- Figure 3.13 animated [p.52] reveals the internal structure of the mitochondrion.
- Figure 3.15 animated [p.53] shows the internal structure of cilia and flagella. (No figure legend in textbook)
- Figure 3.16 animated [p.54] demonstrates a concentration gradient.
- Figure 3.17 animated [p.55] show how the concentration of a solute affects water movement.
- Figure 3.18 animated [p.55] demonstrates the effects of the tonicity of solutions on animal cells.
- Figures 3.19 and 3.20 [pp.56-57] provide a good visual summary of the various types of transport across a cell membrane.
- Figure 3.22 animated [p.58] demonstrates the ATP cycle
- Figure 3.23 animated [p.59] visualizes the function of an enzyme's active site.
- Figure 3.24 animated [p.60] demonstrates the steps of glycolysis.
- Figure 3.25 animated [p.61] demonstrates the formation of ATP by the electron transport system.
- Figure 3.26 animated [p.62] summarizes the process of aerobic cellular respiration.

INTERACTIVE EXERCISES

CHAPTER INTRODUCTION [p.41]

3.1. WHAT IS A CELL? [pp.42-43]

3.2. ORGANELLES OF A EUKARYOTIC CELL [p.44]

Boldfaced Terms

cell theory_____

_____

plasma membrane _____

_____

cytoplasm_____

_____

cytosol _____

_____

prokaryotic cell _____

_____

eukaryotic cell _____

_____

organelle _____

_____

surface-to-volume ratio_____

_____

lipid bilayer _____

_____

Short Answer

1. List the three basic principles of the cell theory. [p.42]

    _____

    _____

    _____

Matching

Match each of the basic components of a cell with its description. [p.42]

2. _____ DNA
3. _____ cytoplasm
4. _____ plasma membrane

A. Everything between the plasma membrane and the region of DNA

B. Outer covering enclosing a cell's internal parts

C. Contains inherited genetic instructions

Short Answer

5. Distinguish prokaryotic cells from eukaryotic cells. [p.42] _____

_____

_____

_____

Dichotomous Choice

Circle one of the two possible answers given between parentheses in each statement. [p.43]

6. Most cells have a (large / small) surface area relative to their volume.

7. Cells require (a lot of / very little) surface area relative to their volumes in order to have enough membrane to take in nutrients and eliminate wastes.

8. As a cell gets bigger, its surface-to-volume ratio (increases / decreases).

9. Surface-to-volume ratio can be increased by a cell being (thin / thick) or by having lots of (smooth surfaces / folded surfaces).

10. The phosphate heads of a phospholipid molecule are (hydrophilic / hydrophobic) while the fatty acid tails are (hydrophilic / hydrophobic).

Labeling

Identify the cellular structures on the cell diagrammed below. [p.44]

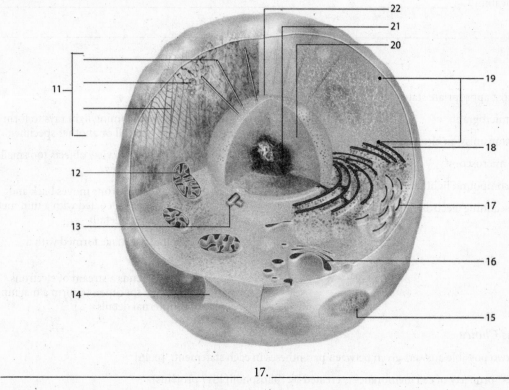

11. _____   17. _____

12. _____   18. _____

13. _____   19. _____

14. _____   20. _____

15. _____   21. _____

16. _____   22. _____

3.3. HOW DO WE SEE CELLS? [p.45]

3.4. THE PLASMA MEMBRANE: A DOUBLE LAYER OF LIPIDS [pp.46-47]

3.5. A WATERY DISASTER FOR CELLS [p.47]

Boldfaced Words

microscopy _____

_____

micrograph _____

_____

selective permeability _____

_____

Matching

Choose the most appropriate statement for each term relating to microscopes. [p.45]

1. ____ micrograph
2. ____ transmission electron microscope
3. ____ microscopy
4. ____ compound light microscope
5. ____ scanning electron microscope

A. Glass lenses bend incoming light rays to form an enlarged image of a cell or another specimen

B. Use of a microscope to view objects too small to be seen by the unaided eye

C. A narrow beam of electrons moves back and forth across a specimen coated with a thin metal layer to reveal surface details

D. A photograph of an image formed with a microscope

E. A magnetic field bends a stream of electrons passing through a specimen to form a magnified image of its internal details

Dichotomous Choice

Circle one of two possible answers given between parentheses in each statement. [p.46]

6. The "fluid" property of cell membranes is created by (phospholipids / proteins).

7. Most cell membrane functions are carried out by (phospholipids / proteins).

Matching

Match the cell membrane component with its description on the right. [p. 46-47]

8. ____ lipid bilayer
9. ____ cholesterol
10. ____ transporter protein
11. ____ receptor protein
12. ____ recognition protein
13. ____ pump protein

A. Moves ions across the membrane using ATP energy

B. Moves substances across the membrane

C. A "sandwich" of phospholipids

D. "Fingerprints" that identify the type of cell

E. Docks for signaling molecules

F. A lipid found in the membranes of human and other animal cells

Fill-in-the Blanks [pp.46-47]

The plasma membrane is a (14) _____ of proteins and lipids. It is considered fluid because most (15) _____ spin, move sideways and flex their tails. By allowing some substances, but not others, to enter and leave a cell, the phospholipids and proteins give the membrane (16) _____. Lipids allow mostly (17) _____ molecules to slip across, while large (18) _____ molecules cross the membrane through the interior of transport proteins. In areas where public sanitation is poor, the bacterium *Vibrio cholerae* causes the disease (19) _____ . The bacterium produces (a)an (20) _____ a poison that affects (21) _____ in the plasma membrane of cells in the (22) _____ . The toxin causes these cells to pump out chloride and sodium ions. Consequently, these cells lose their (23) _____ by osmosis. Cholera's main symptom is massive water loss through (24) _____ .

3.6. THE NUCLEUS [pp.48-49]

3.7. THE ENDOMEMBRANE SYSTEM [pp.50-51]

3.8. MITOCHONDRIA: THE CELL'S ENERGY FACTORIES [p.52]

3.9. THE CELL'S SKELETON [p.53]

Boldfaced Terms

nucleus _____

_____

nuclear envelope _____

_____

nucleolus _____

_____

chromatin _____

_____

chromosome _____

_____

endomembrane system _____

_____

endoplasmic reticulum (ER) _____

_____

ribosome _____

_____

Golgi body _____

_____

vesicle_____

_____

lysosome_____

_____

peroxisomes_____

_____

mitochondrion_____

_____

cytoskeleton_____

_____

microtubules_____

_____

microfilaments_____

_____

intermediate filaments_____

_____

cilia_____

_____

flagella_____

_____

centrioles_____

_____

Matching

Select the description that best matches the nuclear-associated structure listed. [pp.48-49]

1. _____ nucleoplasm
2. _____ nucleolus
3. _____ chromatin

4. _____ nuclear envelope
5. _____ chromosome

A. All the DNA molecules and their attached proteins

B. Fluid part of the nucleus

C. Site of production of ribosomes

D. Section of DNA molecule with associated proteins

E. Double membrane between nucleus and cytoplasm

Matching

Select the function that most closely fits each structure associated with the endomembrane system. [pp.50-51]

6. _____ lysosome

7. _____ Golgi body

8. _____ smooth endoplasmic reticulum (ER)

9. _____ rough endoplasmic reticulum (ER)

10. _____ peroxisome

11. _____ ribosomes

A. Assembles proteins from amino acids

B. Covered with ribosomes; modifies new polypeptide chains

C. Vesicles that break down fatty acids, amino acids, or alcohol

D. Sac of digestive enzymes; digests cell or cell parts, recycles materials

E. Synthesizes lipids; inactivate drugs and toxins

F. Finishes, sorts, and ships proteins, lipids, and enzymes

Sequence

Place the numbers 1-6 in the spaces before each statement to put into proper sequence these events in the organelles of the endomembrane system. [pp.50-51]

12. _____ Polypeptide chains are modified inside the channels of rough ER

13. _____ Ribosomes synthesized in the nucleolus become attached to the ER

14. _____ Golgi body adds the finishing touches to a protein and packages it for transport

15. _____ Ribosomes synthesize polypeptide chains from amino acids

16. _____ A transport vesicle leaves the Golgi and merges with the plasma membrane

17. _____ A transport vesicle leaves the ER and travels to the Golgi body

Choice

Select the cellular component from the following choices which most closely fits each function. [pp.52-53]

mitochondrion   cytoskeleton   "9+2 array"   centriole

18. _____ Internal structure of cilia and flagella.

19. _____ The system of interconnected fibers and threads in the cytosol.

20. _____ Internal structure contains mtDNA, ribosomes and folded inner membrane

21. _____ Composed of microtubules, microfilaments, and intermediate filaments

22. _____ Produces ATP by aerobic cellular respiration

23. _____ "Basal body" found at origin of cilia and flagella

24. _____ Has an important role in cell division

3.10. HOW DIFFUSION AND OSMOSIS MOVE SUBSTANCES ACROSS MEMBRANES [pp.54-55]

3.11. OTHER WAYS SUBSTANCES CROSS CELL MEMBRANES [p.56]

Boldfaced Terms

concentration gradient_____

_____

diffusion_____

_____

passive transport _____

_____

osmosis _____

_____

isotonic _____

_____

hypotonic_____

_____

hypertonic _____

_____

facilitated diffusion _____

_____

active transport _____

_____

endocytosis_____

_____

phagocytosis _____

_____

exocytosis_____

_____

Matching

Match the term to the definition. [pp.54-57]

1. ____ osmosis
2. ____ passive transport
3. ____ endocytosis
4. ____ selective permeability
5. ____ active transport
6. ____ diffusion
7. ____ electric gradient
8. ____ exocytosis
9. ____ tonicity
10. ____ facilitated diffusion
11. ____ phagocytosis
12. ____ concentration gradient

A. Solutes diffuse across a plasma membrane through a transport protein down their own concentration gradients

B. Property of allowing some substances but not others to cross a plasma membrane

C. Net movement of like molecules or ions from an area of higher concentration to an area of lower concentration

D. Difference in electric charge across a plasma membrane

E. Diffusion of water across a selectively permeable membrane in response to solute concentration gradients

F. Refers to the relative solute concentrations in two fluids

G. Solutes are made to cross plasma membranes against their own concentration gradients

H. A small section of plasma membrane folds inward and pinches organic matter from outside the cell into a vesicle; "cell eating"

I. A vesicle moves to the plasma membrane and fuses with it, releasing its contents to the outside of the cell

J. Difference in the number of molecules or ions of a given substance in two neighboring regions

K. Literally means "coming inside the cell"

L. Movement of solutes that does not require energy from ATP

True/False

If the statement is true, write a "T" in the blank. If the statement is false, write an "F" in the blank. [pp.54-57]

13. _____ The rate of diffusion is slower when the concentration gradient is steep.

14. _____ In a solution with more than one solute, each solute diffuses separately from others, according to its own concentration gradient.

15. _____ Osmosis is the diffusion of water across a selectively permeable membrane in response to solute concentration gradients.

Cells and How They Work   **41**

16. _____ If an animal cell were placed in a hypertonic solution, it would swell and perhaps burst.

17. _____ Physiological saline is 0.9 percent NaCl; red blood cells placed in such a solution will not gain or lose water; therefore, one could state that the fluid in red blood cells is hypertonic to the physiological saline.

18. _____ Cell membranes display selective permeability.

19. _____ Red blood cells shrivel and shrink when placed in a hypotonic solution.

20. _____ Exocytosis occurs when a cell takes in a substance by engulfing it into a vesicle derived from the plasma membrane.

Short Answer

21. Distinguish between the terms hypotonic, hypertonic and isotonic. [p.55] _____

_____

_____

3.12. WHEN MITCHONDRIA FAIL [p.57]

3.13. METABOLISM: DOING CELLULAR WORK [pp.58-59]

Boldfaced Terms

metabolism _____

_____

ATP/ADP cycle _____

_____

anabolism _____

_____

catabolism _____

_____

substrates _____

_____

active site _____

_____

Matching

Match the term to the definition [pp.58-59]

1. _____ active site
2. _____ catabolism
3. _____ ATP/ADP cycle
4. _____ metabolic pathway
5. _____ anabolism
6. _____ substrate
7. _____ intermediate
8. _____ phosphate groups
9. _____ enzyme
10. _____ ATP

A. Catalytic molecule that speeds up the rate of a specific chemical reaction

B. $ADP + P_i \rightarrow ATP$; $ATP \rightarrow ADP + P_i$

C. A surface crevice on an enzyme where it interacts with a substrate

D. Pathway where large molecules are broken down into products that have less energy

E. Modified nucleotide that transfers energy from reactions releasing it to those requiring it

F. Any substance that forms between the start and end of a pathway

G. A series of reactions occurring in orderly steps that are catalyzed by enzymes

H. Any substance that enters an enzyme-catalyzed reaction; also called a reactant or a precursor

I. Pathway in which small molecules are put together into more complex molecules

J. Energy of ATP is in the bond between the second and third of these

True/False

Write a T or an F in front of each statement below. [pp.57-59]

11. _____ Nearly all enzymes are proteins.

12. _____ Catalysts are substances that produce energy for anabolic reactions.

13. _____ Enzymes can be used many times by a cell.

14. _____ An enzyme can interact with many different types of molecules.

15. _____ The higher the temperature rises, the faster enzymes will work.

16. _____ Most enzymes in our bodies work best in a pH range of 7.35 to 7.4.

17. _____ Vitamins are a source of coenzymes.

18. _____ Luft's syndrome is linked to defective Golgi bodies.

Matching

Match the items on the following sketch with the correct description. [p.59]

19. ____
20. ____
21. ____
22. ____

A. Substrate molecules
B. Active site
C. Enzyme
D. Product

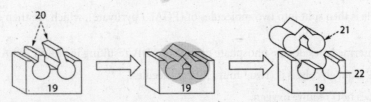

3.14. HOW CELLS MAKE ATP [pp.60-61]

3.15. SUMMARY OF CELLULAR RESPIRATION [p.62]

3.16. OTHER ENERGY SOURCES [p.63]

3.17. NO THANKS TO ARSENIC [p.63]

Boldfaced Terms

cellular respiration _____

_____

glycolysis _____

_____

phosphorylation_____

_____

Krebs cycle _____

_____

electron transport system _____

_____

Dichotomous Choice

Circle one of two possible answers given between parentheses in each statement. [pp.60-62]

1. Cells make ATP by breaking bonds in (nucleic acids / carbohydrates) especially, but also proteins and lipids.

2. During breakdown reactions, energy associated with (calories / electrons) drives ATP formation.

3. Cells of the human body typically form ATP by cellular respiration which is an (aerobic / anaerobic) process.

4. Aerobic pathways start with a set of reactions known as (electron transport / glycolysis).

5. The most common raw material for cellular respiration is (fat / glucose).

6. Glycolysis occurs in the (mitochondria / cytoplasm).

7. For every glucose molecule entering glycolysis, (two / four) three-carbon pyruvate molecule are produced.

8. The first steps of glycolysis are (energy-releasing / energy-requiring).

9. The first steps of glycolysis proceed only when two ATP molecules each transfer a phosphate group to (pyruvate / glucose), donating energy to it in a process called phosphorylation.

10. The glucose molecule is then split into two molecules of (PGAL / pyruvate), which are then converted to an intermediate.

11. During glycolysis, intermediates donate phosphate groups to ADP, resulting in (four / 36) ATP molecules.

12. The net yield of ATP from glycolysis is (two / four) ATP molecules.

13. Glycolysis (does / does not) require oxygen.

44 Chapter Three

14. Following glycolysis, the two pyruvates enter (the cytoplasm / a mitochondrion).

15. In the preparatory steps, a carbon is removed from each pyruvate, leaving a fragment that is combined with (coenzyme A / carbon dioxide), resulting in acetyl-coA.

16. The two-carbon fragment resulting from the preparatory steps is transferred to the (Krebs cycle / electron transport chain).

17. The carbons in the two pyruvates are lost in the Krebs cycle as (NADH / CO_2).

18. Reactions during the Krebs cycle produce (two / 36) molecules of ATP directly.

19. Reactions prior to and during the Krebs cycle produce a large number of (phosphate / coenzyme) molecules such as NADH and $FADH_2$.

20. $FADH_2$ and NADH molecules, loaded with energy removed from glucose, are used in the (Krebs cycle / electron transport system).

21. Electron transport systems and neighboring transport proteins are embedded in the (outer / inner) membrane that divides the mitochondrion into two compartments.

22. An H^+ concentration forms in the outer compartment of the mitochondrion after the pumping of H^+ from the inner compartment; H^+ then flows back into the inner compartment, where (ADP / ATP) is formed.

23. At the end of the electron transport systems, oxygen withdraws electrons and then combines with H^+, resulting in the formation of (carbon dioxide / water).

24. For each molecule of glucose entering the aerobic pathway including glycolysis, (32 / 36) total molecules of ATP are formed.

Short Answer

25. How does arsenic work to poison cells? [p.63] _____

_____

_____

_____

26. Compare lactate fermentation in skeletal muscle cells to aerobic cellular respiration in regards to oxygen requirement. [pp.62-63] _____

_____

_____

_____

27. What is the storage sugar found in humans? Where is it stored in the body? [p.63] _____

_____

_____

_____

28. When are triglycerides used as an energy source? In what tissue is most of the body's fat stored? [p.63] _____

_____

_____

_____

29. What happens to excess proteins in the body? [p.63] _____

_____

_____

_____

30. Why do muscles feel sore after sudden, intense exercise? [p.63] _____

_____

_____

_____

SELF-QUIZ

___ 1. Which of the following is not found in a prokaryotic cell? [p.42]

    a. DNA
    b. plasma membrane
    c. cytoplasm
    d. nucleus

___ 2. Which of the following is not one of the three fundamental features of all cells? [p.42]

    a. cell wall
    b. plasma membrane
    c. cytoplasm
    d. DNA

___ 3. In a lipid bilayer, phospholipid tails point inward and form a(n) _____ layer. [p.43]

    a. acidic
    b. hydrophilic
    c. basic
    d. hydrophobic

___ 4. The _____ creates an image by passing a beam of electrons back and forth across the surface of a specimen coated with a thin layer of metal. [p.45]

    a. compound light microscope
    b. scanning tunneling microscope
    c. transmission electron microscope
    d. scanning electron microscope

___ 5. Which of the following membrane proteins moves molecules by active transport? [pp.46-47]

    a. receptor proteins
    b. recognition proteins
    c. transporter proteins
    d. pump proteins

___ 6. The cellular structure in animals that is involved in the process of intracellular digestion is the _____. [p.51]

    a. lysosome
    b. rough endoplasmic reticulum
    c. microtubules
    d. mitochondria

___ 7. The nucleolus is the site where _____. [p.48]

    a. the protein and RNA subunits of ribosomes are assembled
    b. the chromatin is formed
    c. chromosomes are bound to the inside of the nuclear envelope
    d. chromosomes duplicate themselves prior to cell division

46 Chapter Three

___ 8. The ____ is free of ribosomes, curves through the cytoplasm, and is the main site of lipid synthesis. [p.50]

   a. lysosome
   b. Golgi body
   c. smooth ER
   d. rough ER

___ 9. Mitochondria convert energy obtained from ____ to forms that the cell can use, principally ATP. [p.52, p.60]

   a. water
   b. glucose and other organic molecules
   c. $NADPH_2$
   d. carbon dioxide

___ 10. The component of the cell that is involved with cellular movement is the _____. [p.53]

   a. nucleolus
   b. cell wall
   c. chromatin
   d. cytoskeleton

___ 11. O_2 and CO_2 and other small, nonpolar molecules move across the plasma membrane by _____. [p.54]

   a. facilitated diffusion
   b. endocytosis
   c. diffusion
   d. active transport

___ 12. Ions and small, water-soluble molecules cross the cell membrane through transport proteins down their concentration gradients by _____. [p.56]

   a. facilitated diffusion
   b. endocytosis
   c. simple diffusion
   d. active transport

___ 13. Phagocytosis is a type of _____ used by cells to engulf organic matter outside the cell. [p.57]

   a. exocytosis
   b. passive transport
   c. endocytosis
   d. hydrolysis

___ 14. Which of the following statements about enzymes is incorrect? [p.59]

   a. Enzymes act as catalysts of chemical reactions.
   b. Enzymes are used up during chemical reactions.
   c. Enzymes interact with substrates at the active site.
   d. Enzymes can only interact with specific kinds of molecules.

___ 15. The final electron acceptor at the end of aerobic respiration is ____. [p.61]

   a. NADH
   b. carbon dioxide (CO_2)
   c. ATP
   d. oxygen

___ 16. When glucose is used as an energy source, the largest amount of ATP is generated by the ____ portion of the cellular respiration. [p.62]

   a. glycolysis
   b. phosphorylation
   c. Krebs cycle
   d. electron transport system

___ 17. Excess sugar is stored in muscle and liver cells as _____. [p.63]

   a. triglycerides
   b. starch
   c. glycogen
   d. insulin

___ 18. The body does not store excess _____. [p.63]

   a. fat
   b. glucose
   c. proteins
   d. carbohydrates

CHAPTER OBJECTIVES/REVIEW QUESTIONS

1. List the three generalizations that together constitute the cell theory. [p.42]

2. Name and describe the three basic parts of all living cells. [p.42]

3. Describe and contrast the distinguishing features of prokaryotic and eukaryotic cells. [p.42]

4. Phospholipid molecules have _____ heads and _____ tails. [p.43]

Cells and How They Work   47

5. _____ isolate chemical reactions and separate sequential chemical activities in the cytoplasm. [p.44]

6. Describe the structure of a cell membrane. [p.46-47]

7. State the functions of the following basic eukaryotic organelles and structures: nucleus, nucleolus, nuclear envelope, chromosomes, chromatin, endoplasmic reticulum (rough and smooth), Golgi bodies, lysosomes, peroxisomes, vesicles, mitochondria, ribosomes, cytoskeleton, microtubules, microfilaments, intermediate filaments, flagellum, cilium, centrioles, and basal body. [pp.48–53]

8. The outermost part of the nucleus, the nuclear _____, has two lipid bilayers. [p.48]

9. Ribosomes subunits are assembled in the _____. [p.48]

10. A DNA molecule and its associated proteins is a(n) _____. [p.49]

11. Explain how the ER, Golgi bodies, and certain vesicles function as the endomembrane system. [pp.50-51]

12. _____ are enzyme sacs that break down fatty acids and amino acids. [p.51]

13. _____ are the ATP-producing powerhouses of eukaryotic cells. [p.52]

14. The _____ gives cells their shape and internal organization. [p.53]

15. When molecules move down a concentration gradient, they move from a region where they are _____ concentrated to a region where they are _____ concentrated. [p.54]

16. What two conditions influence the rate of diffusion of charged atoms and molecules? [p.54]

17. Compare and contrast the processes of diffusion and osmosis. [pp.54–55]

18. Two fluids are said to be _____ when solute concentrations are equal on both sides of a cell membrane; when solute concentrations are unequal, one fluid is _____ (fewer solutes) and the other is _____ (more solutes). [p.55]

19. Describe the mechanisms involved in active transport and facilitated diffusion. [p.56]

20. Explain the role of vesicles during exocytosis and endocytosis. [pp.56-57]

21. All of the chemical activity occurring in a cell is called _____. [p.58]

22. Distinguish between biosynthetic metabolic pathways (anabolism) and degradative metabolic pathways (catabolism). [p.58]

23. What is a metabolic pathway? What is necessary for each step in the pathway to occur? [pp.58-59]

24. Define and state the role of each of the following participants in a metabolic pathway: substrate, intermediate, enzyme, coenzyme, energy carrier, and product. [p.59]

25. _____ are molecules that enzymes can chemically recognize, bind and modify. [p.59]

26. Substrates interact with enzymes at the _____ sites of the enzymes. [p.59]

27. The main energy carrier in cells is _____. [p.60]

28. What is the role of NAD^+ in ATP production? [p.61]

29. Where in a eukaryotic cell does glycolysis occur? the Krebs cycle? the electron transport system? [pp.60-61]

30. Describe how ATP is produced with a transport mechanism across the inner membrane of the mitochondrion. [p.61]

31. List (in order) and cite the highlights of the three major stages of aerobic respiration. [p.62]

32. Describe how alternative energy sources such as fats and proteins may enter the energy-releasing pathways. [p.63]

INTEGRATING AND APPLYING KEY CONCEPTS

1. The electron transport system moves electrons across the inner membrane of a mitochondrion by active transport. Active transport requires energy. What is the source of energy for the electron transport system? What is the payback for using this energy?

2. Exactly how does being deprived of oxygen, such as being suffocated with a pillow, kill a person?

3. Since chemical reactions in a cell or organism rely on enzymes, what would be the consequences of enzyme malfunction due an inherited genetic enzyme mutation that affected the shape of the active site?

4. What is the challenge to an aquarium owner to keep marine tropical fish in a well balanced salt-water tank and fresh water tropical fish in a well balanced fresh-water aquarium? (Think tonicity and osmosis.)

4

TISSUES, ORGANS AND ORGAN SYSTEMS

INTRODUCTION

Cells in multicellular organisms face the challenge of job specialization while maintaining the unity and function of the entire organism. This requires cooperation and communication. Cells with similar structure and function organize themselves into tissues. Different tissue types come together to cooperate in the function of an organ. Organs with similar functions work together in organ systems to achieve a common goal. This chapter explores the relationships among tissues, organs and organ systems. You will learn what happens at junctions where cells touch. Also included is a look at the body cavities in which organs are found and the membranes that line them. You will learn that homeostasis is a dynamic balance to which all cells, tissues and organs contribute. Mastering a knowledge of tissue structure and function is essential to the study of individual organ systems which follows in later chapters. The concept of homeostatic feedback mechanisms is a key to understanding the ability of a multicellular organism to function as a whole.

FOCAL POINTS

- Table 4.1 [p.69] and Figure 4.1, animated, [p.68] provide a good review of epithelial tissues
- Figure 4.3 animated [pp.70-71] and Table 4.2 [p.70] provide a good review of connective tissue types
- Figure 4.5 animated [p.72] shows the various types of muscle tissue
- Figure 4.7 animated [p.74] demonstrates the three basic types of cell junctions
- Figure 4.9 animated [p.76] shows the interrelationships among tissues, organs and organ systems
- Figure 4.10 animated [p.77] provides an overview of the body's eleven organ systems
- Figure 4.11 animated [p.78] details the structure of the skin
- Figure 4.13 [p.80] and Figure 4.14 [p.81] shows the three basic components of a negative feedback mechanism
- Figure4.15 animated [p.82] illustrates the mechanism of body temperature regulation

INTERACTIVE EXERCISES

CHAPTER INTRODUCTION [p.67]

4.1. EPITHELIUM: THE BODY'S COVERING AND LININGS [pp.68-69]

Boldfaced Terms

tissue _____

_____

epithelium _____

_____

basement membrane _____

_____

gland _____

_____

exocrine gland _____

_____

endocrine gland _____

_____

True/False

Place a T or F in the blank beside each statement. [p.67]

1. _____ Stem cells are undifferentiated cells that can develop into many other cell types.

2. _____ Stem cells harvested from adults produce more tissue types than those from embryos.

3. _____ Stem cells are the first to form when a fertilized egg starts dividing.

4. _____ Adult stem cells may be able to regenerate cartilage and heart muscle.

5. _____ Stem cell therapy is so promising that all scientists and nonscientists believe in embryonic stem cell research.

Fill-in-the-Blanks [pp.68-69]

A tissue is a group of (6) _____ that perform a certain function. Epithelial tissue has a (7)

_____ surface, which faces either the outside environment or a body fluid. The other surface adheres to

a(n) (8) _____ membrane, a noncellular layer packed with proteins and polysaccharides. There are two

basic types of epithelial tissue: (9) _____ epithelium with one cell layer and (10) _____

epithelium with more than one layer of cells. (11) _____ epithelium is simple epithelium that appears

to be more than one layer because the nuclei of neighboring cells don't line up.

The two basic types of epithelial tissue are placed in categories depending on the (12) _____

of cells at the tissue's free surface. Both simple and stratified epithelium may be any of these three types: (13)

_____ , (14) _____ , (15) _____ .

A(an) (16) _____ makes and secretes products such as saliva or mucus. (17)

_____ glands secrete substances onto an epithelial surface through ducts or tubes. (18)

_____ glands do not secrete substances through tubes or ducts. Their products, (19) _____

are secreted directly into the extracellular fluid bathing the glands.

Identification

Name the tissue types seen in the following illustrations. Be sure to include in your answers which of the two basic types of epithelium is shown and to which shape category it belongs. [pp.68-69]

20. _____ _____
21. _____ _____
22. _____ _____

4.2. CONNECTIVE TISSUE: BINDING, SUPPORT, AND OTHER ROLES [pp.70–71]

Boldfaced Terms

connective tissue _____

_____

matrix _____

_____

fibrous connective tissue _____

_____

cartilage _____

_____

bone tissue _____

_____

adipose tissue _____

_____

Fill-in-the-Blanks [pp.70-71]

The most abundant and widely distributed body tissue is (1) _____ tissue. In most types of

connective tissue, the cells secrete fibrous proteins and produce a (2) _____ of polysaccharides. The

ground substance and protein fibers form a (3) _____ around the cells that gives each connective tissue

its properties. Fibrous connective tissue contains fibers made of (4) _____ and (5) _____ in

its matrix. The matrix of (6) _____ consists of collagen and elastin fibers in a rubbery ground substance.

Calcium salts contribute hardness to (7) _____ tissue. (8) _____ tissue stores fat. The

matrix of (9) _____ is a liquid plasma in which protein fibers and cells are found.

Choice

For questions 10-16, choose from the following: [p.70]

    a. loose connective tissue   b. dense connective tissue   c. elastic connective tissue

10. _____ Flexible tissue under skin and epithelia

11. _____ Has only a few fibers and cells in a jellylike matrix

12. _____ Many collagen fibers impart a less flexible, stronger support

13. _____ Many elastin fibers allow organs like lungs to stretch

14. _____ Tendons and ligaments composed of tear-resistant parallel bundles of collagen

15. _____ Helps support the skin, surrounds muscles and organs

16. _____ Reticular form is the framework for soft organs like the liver and spleen

Choice

For questions 17–21, choose from the following: [p.71]

    a. hyaline cartilage   b. elastic cartilage   c. fibrocartilage

17. _____ The most common type of cartilage

18. _____ Thick bundles of collagen form cartilage "cushions" in joints such as the knee and in the disks that separate vertebrae of the spinal column

19. _____ An early embryo's skeleton is made of this

20. _____ Provides a friction-reducing cover at the ends of freely movable mature bones at joints

21. _____ Collagen and elastin fibers impart a flexible yet rigid structure; found in outer flaps of ears

Choice

For questions 22–28, choose from the following: [p.71]

    a. bone   b. adipose   c. blood

22. ____   Stores fat in droplets

23. ____   Hard tissue due to calcium salts

24. ____   Main role is transport

25. ____   Main tissue of the skeleton

26. ____   Plasma forms its liquid matrix

27. ____   Provides insulation and cushioning

28. ____   Most of this tissue is located just beneath the skin

Identification

Identify the tissue types shown in the following illustrations. [pp.70-71]

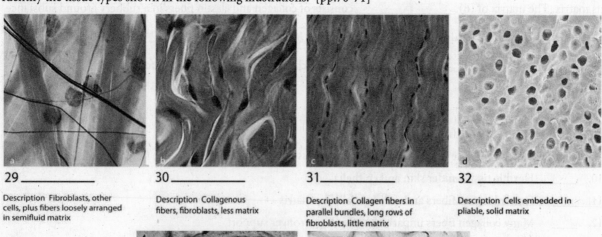

29 _____

Description Fibroblasts, other cells, plus fibers loosely arranged in semifluid matrix

30 _____

Description Collagenous fibers, fibroblasts, less matrix

31 _____

Description Collagen fibers in parallel bundles, long rows of fibroblasts, little matrix

32 _____

Description Cells embedded in pliable, solid matrix

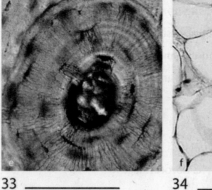

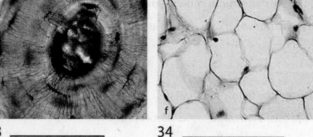

33 _____

Description Collagen fibers, matrix hardened with calcium

34 _____

Description Large, tightly packed fat cells occupying most of matrix

4.3. MUSCLE TISSUE: MOVEMENT [p.72]

4.4. NERVOUS TISSUE: COMMUNICATION [p.73]

4.5. HEALING WITH STEM CELLS AND LAB-GROWN TISSUES [p.73]

Boldfaced Terms

muscle tissue _____

_____

nervous tissue _____

_____

neurons _____

_____

dendrites _____

_____

axons _____

_____

glial cells _____

_____

Identification

Identify the specific type of tissue illustrated below. [pp.72-73]

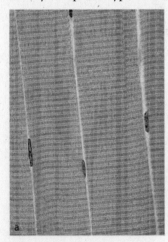

1 _____

Bundles of long, cylindrical, striated
muscle fibers

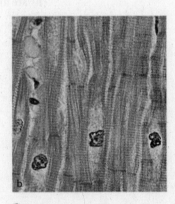

2 _____

Cylindrical muscle fibers with communication
junctions; contract rapidly as a unit

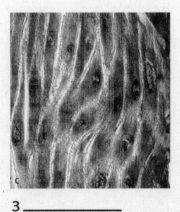

3 _____

Contractile cells tapered at both ends;
involuntarily

Tissues, Organs and Organ Systems   **55**

Choice

For questions 4–11, choose from the following: [p.72]

    a. skeletal   b. smooth   c. cardiac

4. _____ Found in the walls of blood vessels, the stomach, and the intestines

5. _____ Found only in the heart

6. _____ Found in muscles attached to bones

7. _____ Involuntary muscle

8. _____ Contractile cells tapered at both ends

9. _____ Communication junctions allow adjacent cells to contract as a unit

10. _____ Cells with parallel fibers called fascicles

11. _____ Striated muscle

Matching

For questions 12-17, make the best match from the choices in the right hand column [p.73]

12. _____ motor neuron

13. _____ axons

14. _____ Schwann cells

15. _____ dendrites

16. _____ neuroglia

17. _____ neurons

A. provide insulation to nerve cells outside the brain and spinal cord

B. cell processes that receive incoming signals

C. cell processes that conduct outgoing messages

D. nerve cell that carries signals to muscles and glands

E. possess cell bodies and cellular processes called axons and dendrites

F. accessory cells located around neurons; Schwann cells are an example

Fill-in-the-Blanks [p.73]

(18) _____ research may lead to therapies to treat diseases like Parkinson's and muscular dystrophy. Other technologies are focused on growing (19) _____ in the laboratory. Scientists at the National Institutes of Health are using (20) _____ to find a cure for sickle cell anemia. Scientists from the University of Minnesota Medical School are attempting to treat epidermolysis bullosa using stem cells from (21) _____ and (22) _____ to produce normal proteins.

4.6. CELL JUNCTIONS: HOLDING TISSUES TOGETHER [p.74]

4.7. TISSUE MEMBRANES: THIN, SHEETLIKE COVERS [p.75]

Boldfaced Terms

tight junction _____

_____

adhering junction _____

_____

gap junction _____

_____

mucous membrane _____

_____

serous membrane _____

_____

cutaneous membrane _____

_____

synovial membrane _____

_____

Labeling

Most epithelial cells and cells of other tissues adhere strongly to one another by means of specialized attachment sites. Identify each example of cell junctions by entering the correct name in the blank below each sketch. [p.74]

1. _____

2. _____

3. _____

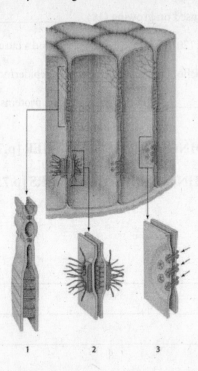

1   2   3

Matching

Refer to the figures in the previous exercise to answer questions 4-6. Match the number corresponding to the cell junction type which best matches the following descriptions. [p.74]

4. _____ Spot-weld junctions holding together the plasma membranes of two adjacent cells; important in tissues subject to stretching, such as epithelium of the skin and stomach.

5. _____ Protein strands forming seals that help stop substances from leaking across a tissue; important in control of what enters the body

6. _____ Channels directly linking the cytoplasm of adjacent cells; allow rapid communication by the direct transfer of ions and small molecules from cell to cell

Complete the Table

Complete the following table by filling in the type of membrane to which each phrase refers. [p.75]

Membrane Type	Description
7.	Occur in paired sheets separated by a thin film of fluid; help anchor and lubricate organs
8.	Secrete fluid to lubricate the ends of moving bones; no epithelial cells present
9.	Hardy, dry membrane; part of the integumentary system
10.	Pink, moist membranes; most have glands that secrete substances such as mucus

Choice

Select the type of membrane that can be found in the following sites in the body. [p.75]

serous   cutaneous   synovial   mucous

11. _____ found where absorption and glandular secretion occur, such as lining the stomach

12. _____ lining capsule-like cavities such as around the knee joint

13. _____ covering the outside of the body

14. _____ lining the digestive, respiratory, urinary, and reproductive tracts

15. _____ enclosing organs such as the heart and lungs

4.8. ORGANS AND ORGAN SYSTEMS [p.76-77]

Boldfaced Terms

organ _____

_____

cranial cavity _____

_____

spinal cavity _____

_____

thoracic cavity _____

_____

abdominal cavity _____

_____

pelvic cavity _____

_____

organ system _____

_____

Sequence

Arrange the following levels of organization in the correct hierarchic order. Find the letter of the simplest level and write it next to the number 1; write the letter of the most complex level next to the number 4, etc. [pp.68,76]

1. ____

2. ____

3. ____

4. ____

A. organ

B. cell

C. tissue

D. organ system

Fill-in-the-Blanks [p.76]

An organ is a combination of two or more kinds of (5) _____ . Many major organs are located in body cavities. The (6) _____ cavity and the (7) _____ cavity house the central nervous system—the brain and spinal cord. The heart and lungs are located in the (8) _____ cavity. A muscular diaphragm separates the thoracic cavity from the (9) _____ cavity, which holds the stomach, liver, most of the intestine, and other organs. Reproductive organs, bladder, and rectum are located within the (10) _____ cavity. Two or more organs make up each of the body's eleven (11) _____

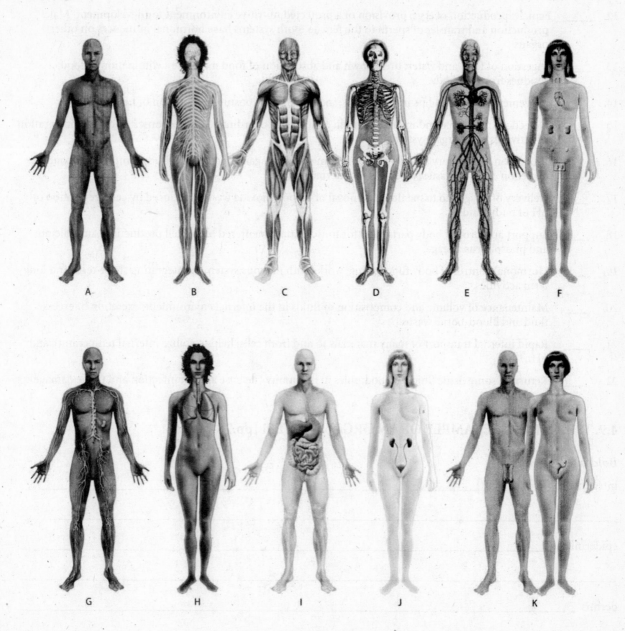

A   B   C   D   E   F

G   H   I   J   K

Match the appropriate function with the letter for each system shown on the accompanying illustrations. [p.77]

12. _____ Female: production of eggs; provision of a protected nutritive environment for development. Male: production and transfer of sperm to the female. Both systems have hormonal influences on other organs.

13. _____ Ingestion of food and water; breakdown and absorption of food molecules; elimination of food residues from the body.

14. _____ Movement of body and its internal parts; maintenance of posture; generation of heat.

15. _____ Detection of external and internal stimuli; control and coordination of responses to stimuli; integration of activities of all organ systems.

16. _____ Protection from injury and dehydration; some defense against pathogens; body temperature control; excretion of some wastes; reception of external stimuli.

17. _____ Delivery of oxygen to tissue fluids; removal of carbon dioxide wastes produced by cells; regulation of pH of body fluids.

18. _____ Support and protect body parts; sites for muscle attachment, red blood cell production, and calcium and phosphorus storage.

19. _____ Hormonal control of body functioning; works with nervous system in integrating short-term and long-term activities.

20. _____ Maintenance of volume and composition of fluids in the internal environment; excretion of excess fluid and blood-borne wastes.

21. _____ Rapid internal transport of many materials to and from cells; helps stabilize internal temperature and pH.

22. _____ Return of some tissue fluid to blood; roles in immunity (defense against infection and tissue damage).

4.9. THE SKIN: AN EXAMPLE OF AN ORGAN SYSTEM [pp.78-79]

Boldfaced Terms

integument _____

_____

epidermis _____

_____

dermis _____

_____

keratinocyte _____

_____

melanocyte _____

_____

blister _____

_____

acne _____

_____

cold sores _____

_____

vitiligo _____

_____

squamous cell carcinoma _____

_____

malignant melanoma _____

_____

Short Answer

1. Name some of the functions of the integument. [pp.78-79] _____

_____

_____

_____

2. Explain the relationship between ultraviolet light and melanocytes. [p.79] _____

_____

_____

_____

3. What is vitiligo? [p.79] _____

_____

_____

_____

Labeling

Label the numbered parts of the accompanying illustration. [p.78]

4. _____

5. _____

6. _____

7. _____

8. _____

9. _____

10. _____

11. _____

12. _____

13. _____

Choice

For questions 14–19, choose from the following: [pp. 78-79]

  a. epidermis   b. dermis   c. hypodermis

14. ____ Fat stored here provides insulation.

15. ____ The stratum corneum is composed of stratified squamous epithelium.

16. ____ Along with the epidermis, this forms the cutaneous membrane.

17. ____ This is the subcutaneous layer.

18. ____ This strong, resilient layer is mainly dense connective tissue.

19. ____ Sweat and oil glands, hair follicles, and nails develop from this layer.

True/False

Answer "True" or "False" for each of the statements that follow. [pp.78-79]

20. _____ Melanocytes are epidermal cells that make carotene.

21. _____ Keratinocytes die by the time they reach the skin surface.

22. _____ Hair follicles and glands are embedded in the hypodermis.

23. _____ Langerhans and Granstein cells have roles in immunity.

24. _____ Keratin makes the stratum corneum tough and waterproof.

25. _____ Fingerprints are different in everyone except identical twins.

Matching

Select the letter of accompanying the following descriptions which best matches the skin disorder associated with it. [p.79]

26. _____ blister

27. _____ acne

28. _____ cold sore

29. _____ squamous cell carcinoma

30. _____ malignant melanoma

A. scabby lesion that indicates an easily treatable form of skin cancer

B. fluid accumulation between the epidermis and dermis

C. a very serious skin cancer with dark, uneven, raised skin lesions

D. bacterial infection of the ducts of oil glands

E. fever blisters caused b a type of Herpes virus

4.10. HOMEOSTASIS: THE BODY IN BALANCE [pp.80-81]

4.11. HOW HOMEOSTATIC FEEDBACK MAINTAINS THE BODY'S CORE TEMPERATURE [p.82-83]

Boldfaced Terms

extracellular fluid _____

_____

sensory receptor _____

_____

integrator _____

_____

effector _____

_____

negative feedback _____

_____

endotherm _____

_____

core temperature _____

_____

hyperthermia _____

_____

hypothermia _____

_____

nonshivering heat production _____

_____

Fill-in-the-Blanks [pp.80-81]

(1) _____ fluid is the fluid outside of cells. Much of this fluid is (2) _____ ,

meaning it occupies spaces between cells and tissues. Cells draw nutrients from it and dump (3) _____

into it.   Drastic changes in its composition can have drastic effects on cell activities. The number and type of (4)

_____ are especially crucial to the maintenance of normal metabolism. (5) _____ means

"staying the same." Homeostatic mechanisms operate to maintain (6) _____ in the volume and

composition of extracellular fluid.

Sensory (7) _____ are cells or cell parts that can detect a(n) (8) _____ , a specific

change in the environment. Your brain is a(n) (9) _____ , a control point where different bits of

information are pulled together in selecting a response. It can send signals to muscles and glands (or both). Muscles

and glands are (10) _____ —they carry out the response. When information from receptors shows that

conditions have deviated from (11) "_____" , the brain functions to bring conditions back within

proper operating range.

Mechanisms for (12) _____ help keep physical and chemical aspects of the body within

tolerable ranges. In (13) _____ feedback, an activity alters a condition in the internal environment, and

this triggers a response that reverses the altered condition. An example is a furnace with a thermostat or the body

temperature of humans being maintained within a normal range. In (14) _____ feedback, a chain of

events is set in motion that intensifies a change from an original condition; childbirth is an example.

Labeling

Identify the numbered components necessary for negative feedback on the diagram below. Enter your answers in the spaces provided. [p.81]

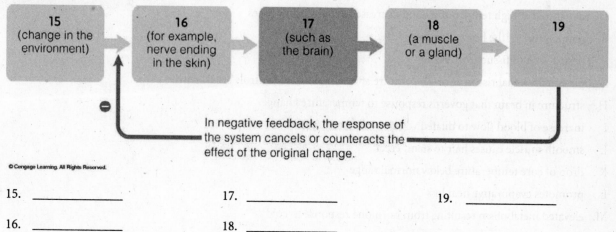

© Cengage Learning. All Rights Reserved.

15. _____

17. _____

19. _____

16. _____

18. _____

Choice

For questions 20–25, choose from the following: [p.81]

      a. negative feedback mechanisms   b. positive feedback mechanisms

20. _____ Set in motion a chain of events that intensify a change from an original condition

21. _____ A furnace with a thermostat

22. _____ After a limited time, the intensifying feedback reverses the change

23. _____ Childbirth

24. _____ Maintaining body temperature in a normal range

25. _____ An activity alters a condition in the internal environment, and this triggers a response that reverses the altered condition

Matching

Write the letter of the correct definition beside each term. [p.82-83]

26. _____ core temperature

27. _____ nonshivering heat production

28. _____ sweating

29. _____ pilomotor response

30. _____ enzymes

31. _____ vasodilation

32. _____ frostbite

33. _____ endotherms

34. _____ hypothermia

35. _____ vasconstriction

36. _____ hypothalamus

37. _____ hyperthermia

38. _____ heat exhaustion

39. _____ heat stroke

A. increase of core temperature above normal range

B. reduction of blood flow to body surface

C. drop in blood pressure due to vasodilation and sweating

D. denatured at high temperatures and decreased activity at low ones

E. temperature of the head and torso

F. destruction of tissues by freezing

G. potentially lethal rise in core temperature after temperature controls break down

H. structure in brain that governs response to temperature change

I. increase of blood flow to dilated vessels at the body surface

J. smooth muscle causes hair to stand erect

K. drop of core temperature below normal range

L. promotes evaporative heat loss

M. elevated metabolism resulting from hormone response to cold

N. organisms that maintain a constant core temperature

SELF-QUIZ

Labeling

Identify each of the following illustrations by labeling it with the appropriate letter from the list of tissues below.

1. ____
2. ____
3. ____
4. ____
5. ____
6. ____
7. ____
8. ____
9. ____
10. ____
11. ____
12. ____
13. ____

A. Adipose [p.71]

B. Bone [p.71]

C. Cardiac muscle [p.72]

D. Dense fibrous connective [p.70]

E. Loose fibrous connective [p.70]

F. Simple columnar epithelium [p. 68]

G. Simple cuboidal epithelium [p. 68]

H. Simple squamous epithelium [p. 68]

I. Smooth muscle [p.72]

J. Skeletal muscle [p.72]

K. Blood [p.71]

L. Cartilage [p70]

M. Nervous [p.73]

 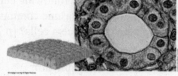 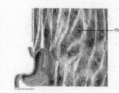 

1.   2.   3.   4.

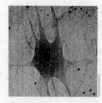 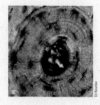 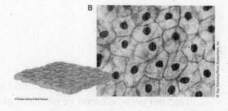

5.   6.   7.   8.

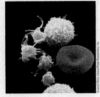

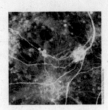

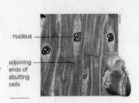

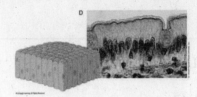

nucleus

adjoining
ends of
abutting
cells

D

9.

10.

11.

12.

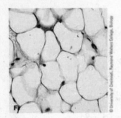

13.

SELF-QUIZ

14. Which of the following stem cells are the most versatile? [p.67]

   a. adult stem cells
   b. bone marrow cells
   c. embryonic stem cells
   d. umbilical cord blood cells

15. All connective tissues contain cells separated by _____ . [p.70]

   a. muscle and nerves
   b. fluid and proteins
   c. fibers and ground substance
   d. blood

16. Which of the following is not a connective tissue? [pp.70–71]

   a. bone
   b. adipose
   c. cartilage
   d. skeletal muscle

17. Which type of muscle cell is found in internal organs, blood vessels and glands? [p.72]

   a. smooth muscle
   b. skeletal muscle
   c. voluntary muscle
   d. cardiac muscle

18. Which of the following cells carry nerve impulses? [p.73]

   a. neuroglia
   b. neurons
   c. Schwann cells
   d. all of the above

19. Which of the following cell junctions allows cytoplasmic communication between adjacent cells? [p.74]

   a. tight junction
   b. adhering junction
   c. gap junction
   d. all of the above

20. Which of the following membranes line the tubes and cavities of the digestive, respiratory, urinary and reproductive systems? [p.75]

   a. serous membrane
   b. cutaneous membrane
   c. synovial membrane
   d. mucous membrane

21. Most cells of the epidermis produce the protein ____; the dermis contains the fibrous proteins ____ and ____. [pp.78-79]

    a. melanin; keratin and collagen
    b. keratin; collagen and elastin
    c. melanin; keratin and hemoglobin
    d. keratin; hemoglobin and melanin

22. Which of the following is **not** found in the dermis? [p.79]

    a. hair follicles
    b. melanocytes
    c. sebaceous glands
    d. sweat glands

23. Which of the following is not associated with a negative feedback mechanism? [pp.80–81]

    a. maintains tolerable ranges of physical and chemical aspects of the body
    b. is similar to the function of a thermostat and furnace
    c. a chain of events intensifies a change from an original condition over a limited period of time
    d. a stimulus triggers a response that reverses the altered condition

24. A significant change in core body temperature can be lethal because it affects _____. [p.82]

    a. sebaceous glands
    b. neuroglia
    c. cardiac muscle
    d. enzymes

CHAPTER OBJECTIVES/REVIEW QUESTIONS

1. Cells are the basic units of life; in a multicellular animal, similar interacting cells and their intercellular substances are grouped into a(n) _____.[p.68]

2. State the structural relationships among cells, tissues, organs, and organ systems. [pp.68-76]

3. _____ always has a free surface, which faces a body cavity or the outside environment. [p.68]

4. _____ glands have no ducts; they release hormones into extracellular fluid. [p.69]

5. Describe the general characteristics of connective tissue and explain how these enable connective tissue to carry out its various tasks. [p.70]

6. Describe the structural differences and functions of loose connective tissue and dense, irregular connective tissue. [pp.70-71]

7. _____ connective tissue is the type found in ligaments and tendons. [p.70]

8. The most common type of cartilage in the human body is _____ cartilage. [p.71]

9. Cartilage in the human ear is _____ cartilage. [p.71]

10. Cartilage that cushions the knee and other joints is _____. [p.71]

11. The specialized function of cells in adipose tissue is for _____ storage. [p.71]

12. Distinguish among skeletal, cardiac, and smooth muscle in terms of location, structure, and function. [p.72]

13. Muscle tissues contain cells that _____ when they receive outside stimulation. [p.72]

14. Neurons can transmit nerve impulses and thus serve as lines of _____. [p.73]

15. Neuron branch processes called _____ pick up incoming chemical messages; others called _____ conduct outgoing messages. [p.73]

16. _____ carry signals called nerve impulses, while _____ support "nerve cells". [p.73]

17. State the general functions of tight junctions, adhering junctions, and gap junctions. [p.74]

18. Various types of thin, sheet-like _____ cover or line organs and may be of the mucous or serous type. [p.75]

19. List the cavities of the human body and generally name the organs found in each. [p.76]

20. List each of the eleven principal organ systems in humans and match each to its main task. [p.77]

21. Explain how keratin in the epidermis protects the rest of your body. [p.78]

22. Describe the two-layered structure of human skin and identify the structures located in each layer. [pp.78-79]

23. Describe the ways by which extracellular fluid helps cells survive. [p.80]

24. Describe the relationships among sensory receptors, integrators, and effectors in a negative feedback system. [pp.80-81]

25. Describe a positive feedback system and give an example. [p.81]

26. Describe homeostatic body responses to a drop or increase in core body temperature. What structure in the brain controls these responses? [pp.82-83]

27. Why is it advantageous for a person who is sweating profusely to drink a sports drink rather than plain water? [p.83]

28. What damaging or lethal events may occur from a serious decrease or increase in core body temperature? [p.83]

29. Explain why fever-reducing drugs may interfere with a strong immune response to an invading pathogen. [p.83]

INTEGRATING AND APPLYING KEY CONCEPTS

1. How does the administration of the labor induction hormone Ptocin (a form of oxytocin) affect the positive feedback mechanism involved in childbirth?

2. If a reduction in the ozone layer of the earth admits more ultraviolet radiation, what parts of the integumentary system will need to respond and what may be the long term consequences for the skin?

3. Consider the effects on the homeostasis of the various organ systems of a marathon runner before, during and after a race.

5

THE SKELETAL SYSTEM

INTRODUCTION

In this chapter, your first goal should be to learn the structure of the bones in the body, and how they are remodeled as we grow, and repaired when they are injured. Next, you should focus on the divisions of the skeleton and which bones are included in each of these divisions as well as the names of the different types of joints. Finally, you should understand how the skeletal system is involved in the homeostasis of all of the body's systems and what may happen if something goes wrong.

FOCAL POINTS

- Figure 5.4 animated [p.91] presents the main bones of the axial and appendicular skeletons. This diagram will be very helpful in learning the names of the bones, especially if you locate the named structures on your own body.
- Figure 5.10 [p.97] illustrates types of movements that our joints allow. These terms are used again in Chapter 6 and other chapters following. This figure will help you to master them now.

INTERACTIVE EXERCISES

CHAPTER INTRODUCTION [p. 87]

5.1. BONE: MINERALIZED CONNECTIVE TISSUE [PP.88-89]

5.2. THE SKELETON: THE BODY'S BONY FRAMEWORK [PP.90-91]

Boldfaced Terms

bone _____

_____

osteoblasts _____

_____

osteocytes _____

_____

compact bone _____

_____

osteon _____

_____

spongy bone _____

_____

epiphysis _____

_____

osteoclasts _____

_____

bone remodeling _____

_____

bone marrow _____

_____

axial skeleton _____

_____

appendicular skeleton _____

_____

ligaments _____

_____

tendons _____

_____

Labeling and Dichotomous Choice

For each numbered part of the diagrams below, choose the correct answer of the two within parentheses.[p.88]

1. a. This outer covering is a two-layer membrane called the (osteon / periosteum).

   b. The outer membrane layer is made of (calcium / dense connective tissue).

   c. The inner layer contains (osteoblasts / compact bone

2. This cavity within a long bone contains (bone marrow / Haversian canals).

3. These fused segments make up (compact / spongy) bone.

4. This dense type of bone is (compact / spongy).

5. Compact bone consists of sets of concentric rings, each called a(n) (osteon / matrix).

6. Within the central canal of each osteon are nerves and (osteoblasts / blood vessels).

7. a. The cells within these cavities are (osteocytes / neurons).

   b. The cavities themselves are (osteoblasts / lacunae).

8. The structure of bone shows that it is (living / dead) tissue.

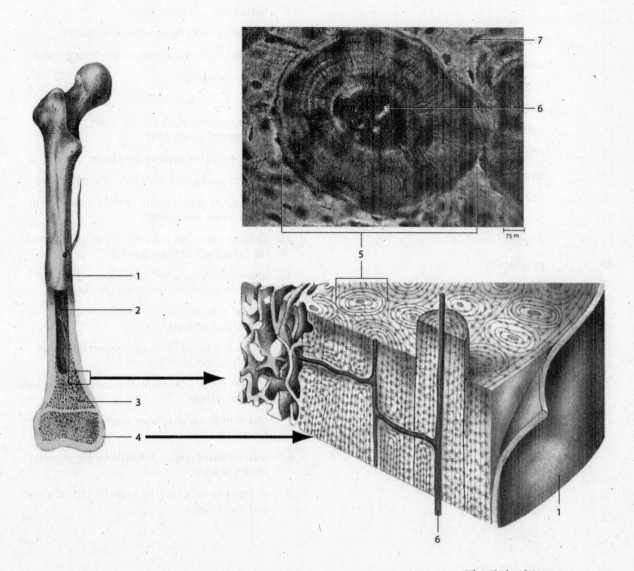

Sequence

9. Indicate the order of the events that occurs in compact bone formation by writing a "1" beside the first event, a "2" by the second, and so on. [p.88]

_____ The cartilage model of the shaft calcifies and blood vessels and nerves infiltrate.

_____ A cartilage "model" of a bone develops in an embryo.

_____ Growth continues in long bones until the epiphyseal plates are replaced by bone

_____ Osteoblasts deposit a bony "collar" that forms the shaft of the bone.

_____ The marrow cavity forms and osteoblasts produce the bone matrix that will calcify.

Matching

Write the letter of the matching definition by the terms below. [pp. 87-89]

10. _____ rubbery skeleton	A.	constant and dynamic depositing and withdrawing of minerals in bone
11. _____ osteoblast	B.	flaring end of a long bone
12. _____ connective	C.	delicate-looking bone inside long bones
13. _____ human growth hormone	D.	cell that secretes collagen in developing bone
14. _____ spongy bone	E.	mature bone cell
15. _____ compact bone	F.	category of tissues to which bone belongs
16. _____ osteoclast	G.	forms the bone shaft and the outer part of the two ends of a long bone
17. _____ osteoporosis	H.	membrane that covers a long bone
18. _____ osteocyte	I.	triggers production of denser, stronger bones
19. _____ lacunae	J.	progressive deterioration of bone, especially in women past menopause
20. _____ collagen	K.	cartilage plate that separates the epiphysis from the bone shaft to allow growth
21. _____ bone remodeling	L.	disorder involving stiff joints due to cartilage breaking down or bone spurs forming
22. _____ osteon	M.	layered compact bone with canal in center (Haversian system)
23. _____ periosteum	N.	connective tissue fibers in compact bone that withstand mechanical stress
24. _____ epiphysis	O.	triggers the development of epiphyseal plates and maintains them
25. _____ epiphyseal plate	P.	spaces in developing bone where osteoblasts are found
26. _____ mechanical stress	Q.	secretes enzymes that break down the organic matrix of bone
27. _____ osteoarthritis	R.	skeleton of early embryo consisting of cartilage and membranes

Fill-in-the-Blank [p. 89]

Calcium is deposited in bone by (28) _____ and is withdrawn when (29) _____

break down the matrix of bone tissue. This bone (30) _____ has several functions. It keeps bone (31)

_____ so it is less likely to break. When a bone is regularly subjected to mechanical stress, more bone is

(32) _____ than is removed. When the body must heal a broken bone, osteoclasts (33)

_____ more calcium than usual from bone matrix. The calcium is used by (34) _____ to

repair the injured bone tissue.

Bone remodeling also plays a role in maintaining the blood level of (35) _____. When the

level falls below the proper range, a hormone called (36) _____ stimulates osteoclasts to break down

bone and release calcium to the blood. If the level of calcium rises too high, the hormone (37) _____

stimulates osteoblasts to deposit calcium in bone tissue. This is an example of (38) _____ feedback.

Short Answer

39. Explain how osteoblasts and osteoclasts work together to increase the diameter of a growing child's bones.
[p.89] _____

_____

_____

_____

40. What are the two types of bone? How do they differ from each other?[p.88] _____

_____

_____

_____

Dichotomous Choice

Circle one of the two possible answers given between parentheses in each statement.[p.90]

41. The long bones of an adult contain mainly (red marrow / yellow marrow).

42. In an adult, most blood cells are formed in (long / irregular) and flat bones.

43. A human has (206 / 260) bones.

44. The body's vertical axis is formed by the (appendicular / axial) skeleton.

45. The bones of the limbs, shoulders and hips make up the (appendicular / axial) skeleton.

46. Bones are connected at joints by (tendons / ligaments).

47. Muscles are held to bone or to each other by (tendons / ligaments).

48. Because of the form in which calcium is deposited into bones, bone tissue is a storage depot for both calcium and (oxygen / phosphorous).

Short Answer

49. List the five functions of bone. [p.90] _____

_____

_____

_____

_____

Identification

Write the scientific name of the numbered bones in the blanks provided. [p.91]

50. _____   68. _____

51. _____   69. _____

52. _____   70. _____

53. _____

54. _____

55. _____

56. _____

57. _____

58. _____

59. _____

60. _____

61. _____

62. _____

63. _____

64. _____

65. _____

66. _____

67. _____

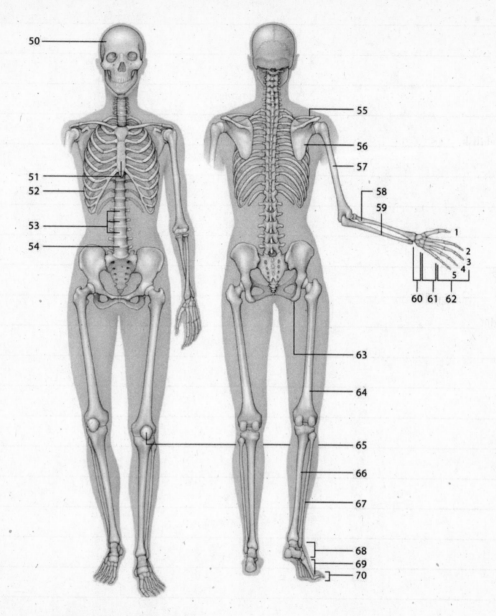

5.3. THE AXIAL SKELETON [PP.92-93]

5.4. THE APPENDICULAR SKELETON [PP.94-95]

Boldfaced Terms

brain case _____

_____

sinuses _____

_____

mandible _____

_____

vertebrae _____

_____

intervertebral disk _____

_____

rib cage _____

_____

sternum _____

_____

pectoral girdle _____

_____

scapula _____

_____

clavicle _____

_____

humerus _____

_____

radius _____

_____

ulna _____

_____

pelvic girdle _____

_____

femur _____

_____

Identification [pp. 92-93]

1. _____ bone

2. _____ bone

3. _____ bone

4. _____ bone

5. _____ bone

6. _____ bone

7. _____

8. _____

9. _____

10. _____

11. _____ vertebrae

12. _____ vertebrae

13. _____ vertebrae

14. _____ (vertebrae)

15. _____ (vertebrae)

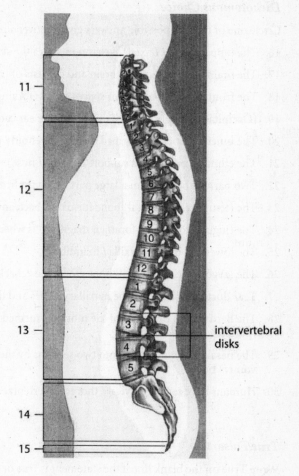

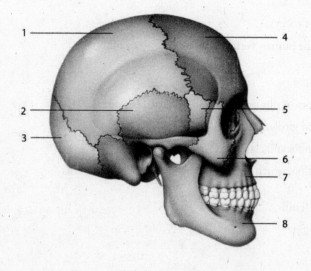

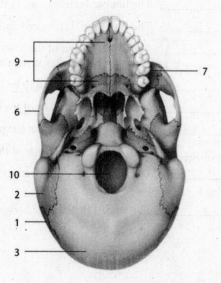

The Skeletal System   **81**

Dichotomous Choice

Circle one of the two possible answers given between parentheses in each statement. [pp.92-93]

16. The (appendicular / axial) skeleton includes the skull, vertebral column, ribs and sternum.

17. The brain case protects the brain and consists of (one / eight) bone(s).

18. The frontal bone contains air spaces called (foramina / sinuses) lined with mucous membranes.

19. (Occipital / Temporal) bones surround the ear canals.

20. The inner eye socket is formed by the (sphenoid / parietal) bone.

21. The ethmoid bone helps support the (jaw / nose).

22. Two parietal bones form a large part of the skull, as does the (zygomatic / occipital) bone.

23. The (temporal / occipital) bone forms the back and base of the skull.

24. The (jugular foramen / foramen magnum) is where the spinal cord emerges from the brain case.

25. The lower jaw is the (maxilla / mandible).

26. The (zygomatic / temporal) bones form the "cheekbones" and outer eye sockets.

27. Tear ducts pass between the maxillary bones and the (palatine / lacrimal) bones.

28. The hard palate, or "roof" of the mouth is formed by extensions of the palatine bones, together with the (parietal / maxillary) bones.

29. The nasal cavity is divided into two sections by the thin nasal septum, formed partially by the (mandible / vomer) bone.

30. Humans have seven (cervical / thoracic) vertebrae.

True/False

Write True on the blank line if the statement is true or False on the blank line if the statement is false. [pp. 93-95]

31. _____ The thoracic vertebrae are located in the chest.

32. _____ The lumbar vertebrae are in the lower back.

33. _____ The "tailbone" is properly called the coccyx.

34. _____ About one-fourth of the length of the human vertebral column consists of tough, compressible lumbar vertebrae.

35. _____ A "slipped" disk is said to be herniated.

36. _____ The "rib cage" consists of twelve pairs of ribs and the paddle-shaped sternum

37. _____ The humerus is a part of the pectoral girdle.

38. _____ The pelvic girdle consists of the scapula and clavicles.

39. _____ The humerus joins the radius and ulna, which join eight small wrist bones known as phalanges.

40. _____ The longest bone in the body is the femur.

5.5. JOINTS: CONNECTONS BETWEEN BONES [PP.96-97]

5.6. DISORDERS OF THE SKELETON [PP.98-99]

Boldfaced Terms

synovial joint _____

_____

cartilaginous joint _____

_____

fibrous joint _____

_____

flexion _____

_____

extension _____

_____

circumduction _____

_____

rotation _____

_____

abduction _____

_____

adduction _____

_____

supination _____

_____

pronation _____

_____

osteoarthritis _____

_____

carpal tunnel syndrome _____

_____

sprain _____

_____

simple fracture _____

_____

complete fracture _____

_____

compound fracture _____

_____

Fill-in-the-Blanks [p. 96]

(1) _____ joints are built to allow movement and are lubricated by (2) _____ fluid

secreted into a capsule made of dense connective tissue that surrounds the bones of the joint. Synovial joints include

(3) _____ joints such as the knee and elbow, as well as (4) _____ joints such as the hip. (5)

_____ joints such as between vertebrae allow only slight movement. An adult's (6) _____

joints, such as those holding teeth in their sockets, don't allow movement.

Identification [p. 97]

7. _____   11. _____

8. _____   12. _____

9. _____   13. _____

10. _____   14. _____

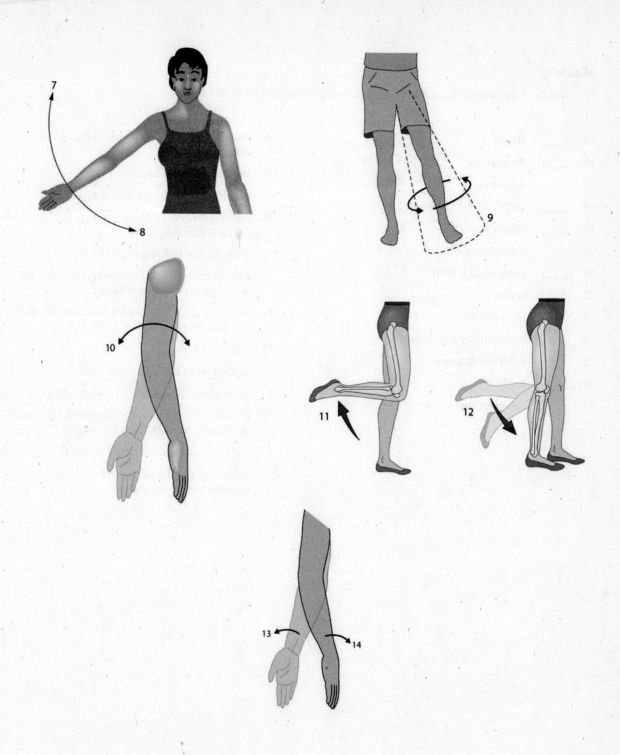

Matching

Match each of the following disorders with its correct definition. [pp.98-99]

15. _____ strain
16. _____ dislocation
17. _____ infection
18. _____ carpel tunnel syndrome
19. _____ simple fracture
20. _____ osteoarthritis
21. _____ compound fracture
22. _____ sprain
23. _____ rheumatoid arthritis
24. _____ osteogenesis imperfecta (OI)
25. _____ complete fracture
26. _____ tendinitis
27. _____ osteosarcoma

A. bacteria from elsewhere may spread to bone via the bloodstream

B. bone is separated into two pieces

C. rapidly spreading cancer of the bone

D. wearing away of cartilage covering the bone ends of freely movable joints

E. stretching or twisting of a joint

F. painful inflammation of tendons between the wrist ligament and wrist bones

G. inherited disorder in which bones break easily due to defective collagen

H. crack in a bone

I. separation of two bones at a joint

J. small tear in ligaments or tendons of a joint

K. erosion of cartilage and inflammation in a joint due to attack by the immune system

L. ends of broken bones puncture the skin

M. inflammation of tendons and synovial membranes around a joint

5.7. THE SKELETAL SYSTEM IN HOMEOSTASIS [P.100]

Matching

Match each of the following systems with its correct definition. [p.100]

1. _____ integumentary

2. _____ muscular

3. _____ digestive

4. _____ cardiovascular/blood

5. _____ immunity/lymphatic

6. _____ respiratory

7. _____ urinary

8. _____ nervous

9. _____ sensory

10. _____ endocrine

11. _____ reproductive

A. The rib cage and sternum protect the lungs.

B. Bone calcium helps this organ pump blood. Bone marrow is used for this.

C. Transmission of special nerve impulses. Skull and facial bones protect organs of this system.

D. The skeleton provides support for this system and the muscles below it.

E. Transmission of all nerve impulses. Skull and vertebrae protect organs of this system.

F. Bones of the rib cage and pelvis protect organs of this system. Teeth used in this system.

G. Organs and glands protected by the pelvis. Milk production in a nursing mother.

H. Bone calcium released for function of components of this system. Used for movement.

I. Calcium released from bones for production and secretion of some hormones.

J. Defense cells of this system form in bone marrow.

K. The rib cage protects a pair of organs of this system while the pelvis protects another organ.

SELF-QUIZ

Choice

For questions 1–10, choose from the following answers:

        a. carpals   b. metatarsals   c. femur   d. humerus   e. ulna   f. tarsals

        g. radius   h. fibula   i. phalanges (used twice)   j. metacarpals   k. tibia

The upper arm bone (1) _____ [p.94] is connected to the lower arm bones (2) _____ and

_____ [p.94]. The lower arm bones are connected to the wrist bones (3) _____ [p.94]. The wrist bones

are connected to the hand bones (4) _____ [p.94]. The hand bones are connected to the finger bones (5)

_____ [p.94].

The upper leg bone (6) _____ [p.95] is connected to the lower leg bones (7) _____ and

_____ [p.95]. The lower leg bones are connected to the anklebones (8) _____ [p.95]. The anklebones are

connected to the foot bones (9) _____ [p.95]. The foot bones are connected to the toe bones (10) _____

[p.95].

Choice

For questions 11–16, choose from the following answers:

    a. spongy bone   b. compact bone   c. Haversian system   d. epiphysis   e. bone marrow   f. bone remodeling

____ 11. Channels for blood vessels and nerves that transport substances to and from osteocytes. [p.88]

____ 12. A connective tissue where blood cells are formed. [p.90]

____ 13. Lacy tissue inside a bone shaft. [p.88]

____ 14. Ongoing calcium recycling in bones. [p.89]

____ 15. Dense tissue that forms a long bone shaft. [p.88]

____ 16. The end of a long bone. [p.88]

For questions 17–20, choose from the following answers: [p.96]

a. synovial   b. cartilaginous   c. fibrous

____ 17. Nonmoving joints (sutures) between skull bones

____ 18. Hips and shoulders

____ 19. Knees and elbows

____ 20. Connections of some ribs to the breastbone

CHAPTER OBJECTIVES/REVIEW QUESTIONS

1. Explain the various roles of osteoblasts, osteoclasts, cartilage models, long bones, and epiphyseal plates in the development of human bones. [pp.88-89]

2. Define osteoporosis and discuss some of its causes and ways to prevent it and identify the individuals most likely to develop this condition. [p.89]

3. Give an example of each of these synovial joint movements: flexion, extension, circumduction, rotation, abduction, adduction, supination, pronation. [p.97]

4. Explain the difference a strain and a sprain. [p.98]

5. Distinguish between the three types of bone fractures. [p.99]

INTEGRATING AND APPLYING KEY CONCEPTS

Explain why, in a child, a fracture at an epiphyseal plate might be more serious than a fracture in the center of a long bone.

6

THE MUSCULAR SYSTEM

INTRODUCTION

The coverage of the muscular system in this chapter has two main themes. These themes of structure and function will be applied to other body systems as they are to muscles in this chapter. Understanding the structure of muscles is essential to understanding how they function. Understanding function clarifies the capabilities and limitations that the structure allows. The relationship of structure and function is one of the most important concepts in Biology. In addition, this chapter will explain the interrelationships between the muscular system and other systems of the body.

FOCAL POINTS

- Figure 6.5 animated [p.107] illustrates the very simple but critical process by which our bodies move. Understanding the opposing action of muscles and muscle groups is necessary before any further investigation into muscle action.
- Figures 6.7 animated [p.108] and 6.9 animated [p.110] are excellent references for mastering terminology used to describe muscle structure and function.
- Figure 6.8 animated [p.109] deserves your concentrated attention. This visual explanation of how muscles contract at the molecular level should be studied as the mechanism by which opposing muscle groups pull the bones and allow coordinated movement.

INTERACTIVE EXERCISES

CHAPTER INTRODUCTION [P. 103]

6.1. THE BODY'S THREE KINDS OF MUSCLE [PP.104-105]

6.2. THE STRUCTURE AND FUNCTION OF SKELETAL MUSCLES [PP.106–107]

6.3. HOW MUSCLES CONTRACT [PP.108–109]

Boldfaced Terms

skeletal muscle _____

_____

smooth muscle _____

_____

cardiac muscle _____

_____

muscular system _____

_____

myofibrils _____

_____

origin _____

_____

insertion _____

_____

sarcomere _____

_____

actin _____

_____

myosin _____

_____

sliding-filament mechanism _____

_____

rigor mortis _____

_____

Complete the Table

Compare the three types of muscle tissue by completing the table below. [p.104]

Muscle Type	Location	*Striated?*	*Voluntary?*
Skeletal	1.	4.	7.
Smooth	2.	5.	8.
Cardiac	3.	6.	9.

Short Answer

10. Of the three muscle types, only skeletal muscle is included in the "muscular system." Why is this type of muscle called "skeletal"? What are its main jobs in the body? [p.104] _____

_____

_____

_____

Identification

Name each numbered muscle on the accompanying illustration. In the parentheses provided, write the letter of its specific movement. [p.105]

11. _____ (   )

12. _____ (   )

13. _____ (   )

14. _____ (   )

15. _____ (   )

16. _____ (   )

17. _____ (   )

18. _____ (   )

19. _____ (   )

20. _____ (   )

21. _____ (   )

22. _____ (   )

23. _____ (   )

24. _____ (   )

25. _____ (   )

26. _____ (   )

A. Bends lower leg at knee when walking, extends foot when jumping

B. Compresses abdomen; helps in lateral rotation of trunk

C. Flexes foot toward the shin

D. Raises arm

E. Bends forearm at elbow

F. Depresses thoracic cavity, compresses abdomen, bends backbone

G. Straightens forearm at elbow

H. Bends thigh at the hip, bends lower leg at knee, rotates thigh in an outward direction

I. Extends and rotates thigh outward when walking, running, climbing

J. Flexes, laterally rotates, and draws thighs toward body

K. Draws arm forward and in toward the body

L. Lifts shoulder blade, braces shoulder, draws head back

M. Draws shoulder blade forward; helps raise arm and push

N. Flexes thigh at the hips, extends leg at the knee

O. Rotates and draws arm backward and toward body

P. Draws thigh backward, bends knee

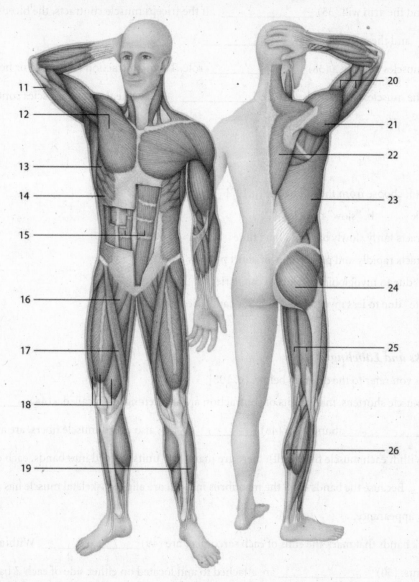

Dichotomous Choice

Circle one of two possible answers given between parentheses in each statement. [p.106]

27. A skeletal muscle contains muscle cells bundled together by (fibrous / connective) tissue.

28. Within each muscle cell are threadlike (myofibrils / tendons).

29. The end of the muscle that is attached to a bone that remains fairly motionless during a movement is the (origin / insertion).

30. The end that is attached to the bone that moves the most is the (origin / insertion).

31. When a skeletal muscle contracts, it (pushes / pulls) on the bones to which it is attached.

Fill-in-the-Blanks [pp. 106-107]

A skeletal muscle contains bundles of cells that look like (32) _____ . Some muscles are

arranged as (33) _____ or groups. For example, if the biceps muscle (34) _____ , the

The Muscular System **93**

triceps will relax and the arm will (35) _____ . If the triceps muscle contracts, the biceps will (36) _____ and the arm will (37) _____ .

Other muscles work in a (38) _____ role. Their contraction adds force or helps stabilize another muscle. The muscles that stabilize the (39) _____ ; while the hand muscles contract are an example.

Choice

For questions 40–43, choose from the following: [p.107]

    a. "fast" muscle   b. "slow" muscle

40. _____   contracts fairly slowly but for a long time

41. _____   contracts rapidly and powerfully for short periods

42. _____   "red" due to myoglobin and lots of capillaries

43. _____   "white" due to less myoglobin and fewer capillaries

Fill-in-the-Blanks and Labeling

Fill in the blanks as you refer to the diagram below. [p. 108]

When a muscle shortens, many units of contraction are shortening, each called a (44) _____.
Within a (45) _____, bundles of (46) _____, also called muscle fibers, are arranged in a parallel manner. Within each muscle fiber (cell), there are many tiny units divided into bands, each of which is a (47) _____. Because the bands of all the myofibrils in a cell are aligned, skeletal muscle has a (48) _____ appearance.

The dark bands that mark the ends of each sarcomere are (49) _____. Within a sarcomere are two sizes of filaments. (50) _____ are attached to and located on either side of each Z band. This region with only thin filaments is the (51) _____. In the very center of each sarcomere, there are only (52) _____ making up the (53) _____. Many thin and thick filaments overlap each other in the (54) _____. mechanism.

Each thin filament is like two strands of pearls twisted together. The "pearls" are (55) _____ molecules. Each thick band is made of molecules of the protein (56) _____ with knoblike heads projecting from each of the bundled molecules.

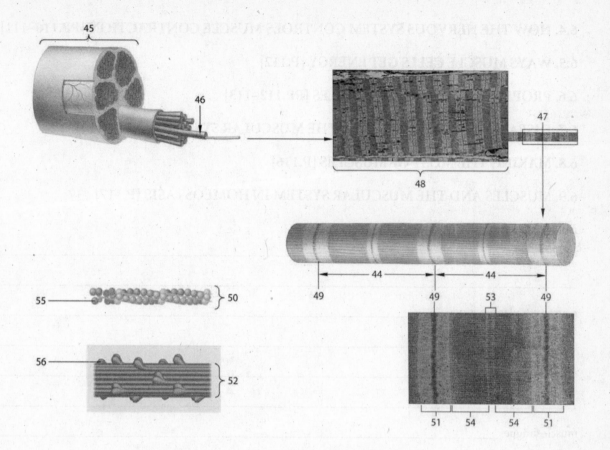

Dichotomous Choice

Circle one of two possible answers given between parentheses in each statement. [p.109]

57. According to the (sarcomere / sliding filament) model, during a contraction all of the (actin / myosin) filaments stay in place.

58. The (actin / myosin) filaments are pulled inward toward the center of the sarcomere.

59. The energy to power the movement is from (DNA / ATP).

60. A change in calcium concentration causes a myosin head to form a (link / cross-bridge) to actin.

61. This link tilts the myosin head and pulls the (actin / myosin) filament inward.

62. The cross-bridge is then broken with the help of (DNA / ATP) and the head returns to its starting position.

63. Each contraction of a sarcomere requires this action to be repeated (one / hundreds of) time(s).

Short Answer

64. What is rigor mortis? What causes it? [p.109] _____

_____

_____

_____

6.4. HOW THE NERVOUS SYSTEM CONTROLS MUSCLE CONTRACTION [PP.110–111]

6.5. WAYS MUSCLE CELLS GET ENERGY [P.112]

6.6. PROPERTIES OF WHOLE MUSCLES [PP.112–113]

6.7. DISEASES AND DISORDERS OF THE MUSCULAR SYSTEM [PP.114-115]

6.8. MAKING THE MOST OF MUSCLES [P.116]

6.9. MUSCLES AND THE MUSCULAR SYSTEM IN HOMEOSTASIS [P.117]

Boldfaced Terms

motor neuron _____

_____

neuromuscular junctions _____

_____

neurotransmitter _____

_____

muscle fatigue _____

_____

oxygen debt _____

_____

motor unit _____

_____

muscle twitch _____

_____

tetanus _____

_____

muscle tone _____

_____

muscle tension _____

_____

muscle spasm _____

_____

muscle cramp _____

_____

muscular dystrophies _____

_____

Duchenne muscular dystrophy _____

_____

myotonic muscular dystrophy _____

_____

botulism _____

_____

tetanus _____

_____

sarcoma _____

_____

aerobic exercise _____

_____

strength training _____

_____

Matching

Match each of the following terms with the appropriate phrase or description. [pp.110-111]

1. _____ calcium
2. _____ T-tubules
3. _____ neuromuscular junction
4. _____ synapse
5. _____ sarcoplasmic reticulum (SR)
6. _____ ACh - acetylcholine
7. _____ troponin

A. place where branched endings of axons interface with muscle cell membranes
B. attaches to actin and tropomyosin
C. small extensions of muscle cell plasma membrane that connect to sarcoplasmic reticulum
D. neurotransmitter that carries a signal from a motor neuron across a synapse to a muscle cell
E. modified endoplasmic reticulum found around myofibrils; releases calcium ions when stimulated by a nerve impulse
F. released from sarcoplasmic reticulum and binds to troponin on actin filament
G. gap between axon ending and plasma membrane of a muscle cell

Sequencing

8. Write the number "1" by the first event to occur in a muscle contraction. Continue numbering the events in the order in which they occur. [pp.110-111]

_____ calcium ions bind to troponin and uncover binding site on actin filament

_____ ACh is released from axon into synapse and diffuses to receptors on muscle cell membrane

_____ SR within muscle cell releases calcium ions that diffuse into myofibrils

_____ nervous system signal is shut off and calcium ions are actively transported into SR

_____ cross-bridges are broken, myosin cannot bind to covered actin sites, muscle relaxes

_____ myosin cross-bridges attach to actin binding site and sarcomere contracts

_____ in the absence of calcium ions, the actin binding site is covered again

_____ muscle neuron carries an impulse to a neuromuscular junction

_____ neural signal spreads over plasma membrane and into T tubules

Sequencing

8. Write the number "1" by the first event to occur in a muscle contraction. Continue numbering the events in the order in which they occur. [pp.110-111]

_____ calcium ions bind to troponin and uncover binding site on actin filament

_____ ACh is released from axon into synapse and diffuses to receptors on muscle cell membrane

_____ SR within muscle cell releases calcium ions that diffuse into myofibrils

_____ nervous system signal is shut off and calcium ions are actively transported into SR

_____ cross-bridges are broken, myosin cannot bind to covered actin sites, muscle relaxes

_____ myosin cross-bridges attach to actin binding site and sarcomere contracts

_____ in the absence of calcium ions, the actin binding site is covered again

_____ muscle neuron carries an impulse to a neuromuscular junction

_____ neural signal spreads over plasma membrane and into T tubules

Choice

For questions 9-15, indicate with which ATP-producing pathway the event is associated. Use these choices. [p.112]

     a. creatine phosphate transfer   b. aerobic respiration   c. glycolysis only

9. ____ muscle fatigue occurs when oxygen debt makes this impossible

10. ____ enzyme transfers phosphate stored in this form to ADP

11. ____ provides most of the ATP for muscle contraction

12. ____ fuels muscle contractions until a slower ATP-forming pathway kicks in

13. ____ occurs when muscle does not receive enough oxygen for aerobic ATP production

14. ____ produces lactic acid which, along with oxygen debt, causes muscle fatigue

15. ____ depends on glucose and fatty acids to fuel ATP production

Matching

Match each of the following terms with the appropriate phrase or description. [pp.112-113]

16. ____ isotonic

17. ____ muscle fatigue

18. ____ muscle tension

19. ____ motor neuron

20. ____ muscle twitch

21. ____ tetanus

22. ____ isometric

23. ____ muscle tone

24. ____ load

25. ____ motor unit

A. force that a contracting muscle exerts on a bone

B. normal sustained contraction occurring when muscle twitches run together

C. steady, low-level contracted state of a muscle

D. contraction of all the cells in a motor unit

E. nerve cell that stimulates a muscle cell to contract

F. the weight of an object, or gravity's pull on a muscle

G. contracting muscles develop tension but don't shorten

H. a motor neuron and the muscle cells it synapses with

I. contracting muscles shorten and move a load

J. muscle can no longer contract

True/False

Write True or False in the blank beside each statement. [pp.112-113]

26. _____ The precision possible for a muscle depends on the number of cells in each motor unit of the muscle.

27. _____ If a new nerve impulse arrives at a muscle before the previous muscle twitch ends, the muscle does not respond to it.

28. _____ A single muscle twitch lasts for about thirty seconds.

29. _____ The bacterial disease tetanus can be fatal because muscles do not relax after a contraction.

30. _____ During muscle tension, all of its cells contract at the same time.

31. _____ When a muscle is not working, all of its cells relax.

32. _____ A muscle shortens when muscle tension exceeds the force of load.

33. _____ When you attempt to lift an object that is too heavy, muscles are contracting isometrically.

34. _____ When a muscle becomes fatigued, it usually recovers and can contract again within a few minutes.

Elimination

For each statement below, cross out the choice that does NOT fit. [p.114]

35. Examples of muscle injury include (tears / cramps / strains).

36. Deficiency of potassium can cause (tics / spasms / cramps).

37. Duchenne muscular dystrophy is (due to a mutation / fatal / a disease of adults).

38. Muscle strains are properly treated with (rest / hot packs / ibuprofen).

39. A sudden involuntary muscle contraction may result in a (cramp / spasm / strain).

40. The effects of a muscle tear may include (scarring / lengthening / reduced function).

Short Answer

41. What do Duchenne muscular dystrophy and myotonic muscular dystrophy have in common? How are they different? [p.114] _____

_____

_____

_____

_____

True/False

Write True or False in the blank beside each statement. [p.116]

42. _____ Prolonged lack of use of a muscle results in atrophy.

43. _____ Strength training increases muscle endurance.

44. _____ Aerobic exercise works muscles at a rate at which the body can keep them supplied with oxygen.

45. _____ Strength training, such as weight lifting, results in muscles that can work longer without becoming fatigued.

46. _____ Fitness experts recommend a training program that includes aerobic workouts and eliminates strength training.

47. _____ Beginning in a person's thirties, the same amount of exercise will not result in as much increase in muscle tension capacity as at a younger age.

Matching

Match each of the following systems with the appropriate contribution of the muscular system to homeostasis. [p.117]

48. _____ integumentary system

49. _____ digestive system

50. _____ immunity and lymphatic system

51. _____ urinary system

52. _____ sensory systems

53. _____ endocrine system

A. Abdominal muscles support the kidneys and bladder.
B. Skeletal muscles support pancreas and thyroid.
C. Skeletal muscles support the skin.
D. Muscles operate in chewing and swallowing.
E. Skeletal muscles move eyes and have sensors for body position.
F. Smooth muscles form the walls of vessels in this system and skeletal muscles help support nodes.

SELF-QUIZ

For questions 1–5, choose from the following answers:

    a. isometric contraction
    b. cross-bridge formation
    c. the sliding-filament mechanism
    d. reciprocal innervation
    e. isotonic contraction

____ 1. Helps coordinate the contraction and relaxation of antagonistic pairs of muscles. [p.106]

____ 2. Muscle shortens and moves a load (e.g., walking downstairs). [p.113]

____ 3. Muscle develops tension; does not shorten during contraction (e.g., holding a glass of lemonade in one position). [p.113]

____ 4. Process assisted by calcium ions and ATP. [p.109]

____ 5. Explains how myosin filaments move to the center of a sarcomere and back. [p.109]

For questions 6-10, choose from the following answers:

    a. actin
    b. myofibril
    c. myosin
    d. sarcomere
    e. sarcoplasmic reticulum

____ 6. Each consists of many repetitive units of muscle contraction; many found in a muscle cell [p.106]

____ 7. Basic unit of muscle contraction [p.108]

____ 8. Thin filaments attached to Z bands [p.108]

____ 9. Stores calcium ions and releases them in response to neural signals [p.110]

____ 10. Thick filaments with double heads [p.108]

CHAPTER OBJECTIVES/REVIEW QUESTIONS

1. Contrast the actions and risks of banned anabolic steroids and human growth hormone used by some athletes and body builders. [p.103]

2. Describe the fine structure of a muscle cell, using the terms *myofibril, sarcomere, actin,* and *myosin.* [pp.106, 108]

3. Explain how skeletal muscles cause movement, including the terms *origin, insertion, bone,* and *antagonistic pairs* in your explanation. [p.106]

4. Compare the three types of muscles found in the body by describing their structures, listing their locations and explaining what they do. [p.104]

5. Define rigor mortis and discuss the time frame associated with and how it is useful for crime investigators. [p.109]

6. Summarize the sliding filament mechanism of muscle contraction. [p.109]

7. List, in sequence, the steps that occur in the signaling between a neuron and a muscle cell. [p.110]

8. Explain the roles of creatine phosphate and cellular respiration during normal muscle activity. [p.112]

9. Explain what occurs in skeletal muscle cells when there is not enough oxygen to allow ATP production by aerobic respiration. [p.112]

10. Distinguish between a spasm, a cramp and a tic. [p.114]

11. Distinguish between Duchenne muscular dystrophy and myotonic muscular dystrophy. [pp.114-115]

12. When does muscle fatigue occur and what factors seem to be involved? [pp. 112-113]

13. Which is the causative agent of botulism food poisoning and how does it cause the disease? [p.115]

INTEGRATING AND APPLYING KEY CONCEPTS

1. Explain how an individual's genetically determined amounts of "fast" versus "slow" muscle cells may influence success at sprinting versus long-distance athletic events.

2. What do creatine phosphate supplements provide to an athlete's muscles? What mode of ATP production would use of this supplement delay?

7

CIRCULATION: THE HEART AND BLOOD VESSELS

INTRODUCTION

As you work your way through this chapter, be aware of your own circulatory system. While you work, it is delivering nutrients and oxygen to all of your cells, as well as removing metabolic wastes and carbon dioxide. This chapter will explain how your heart and blood vessels function, as well as what can occur when they do not work correctly.

FOCAL POINTS

- Figure 7.1 animated [p.122] diagrams the main components of the cardiovascular system. Imagine what this diagram would look like if it included the thousands of miles of tiny capillaries supplying your tissues.
- Figure 7.6 animated [p.125] illustrates the cardiac cycle.
- Figure 7.7 animated [pp.126-127] illustrates the two circulatory routes that blood follows when it leaves the heart. Mastering the sequence of blood flow between heart chambers as well as to and from the lungs and tissues is essential to understanding the circulatory system.

INTERACTIVE EXERCISES

CHAPTER INTRODUCTION [p.121]

7.1. THE CARDIOVASCULAR SYSTEM: MOVING BLOOD THROUGH THE BODY [pp.122-123]

7.2. THE HEART: A MUSCULAR DOUBLE PUMP [pp.124-125]

7.3. THE TWO CIRCUITS OF BLOOD FLOW [pp.126-127]

7.4. HOW CARDIAC MUSCLE CONTRACTS [p. 128]

Boldfaced Terms

cardiovascular system _____

_____

heart _____

_____

arteries _____

_____

arterioles _____

_____

capillaries _____

_____

venules _____

_____

veins _____

_____

myocardium _____

_____

atrium (plural: atria) _____

_____

ventricle _____

_____

atrioventricular (AV) valve _____

_____

pulmonary valve _____

_____

aortic valve _____

_____

coronary circulation _____

_____

aorta _____

_____

cardiac cycle _____

_____

systole _____

_____

diastole _____

_____

pulmonary circuit _____

_____

systemic circuit _____

_____

hepatic portal system _____

_____

cardiac conduction system _____

_____

sinoatrial (SA) node _____

_____

atrioventricular (AV) node _____

_____

Fill-in-the-Blanks [p.123]

The (1) _____ system, also called the circulatory system, is built to circulate blood to every

living cell in the body. The two main elements in the system are the (2) _____ and (3)

_____ , which are tubes of different diameters. The heart pumps blood into large-diameter (4)

_____ . From there the blood flows into smaller (5) _____ which branch into even

narrower (6) _____ . Blood flows from capillaries into small (7) _____ then into large-diameter (8) _____ that return blood to the heart. Blood flows (9) _____ through arteries,

but in capillary beds, it must flow (10) _____ .

        Blood is called the "river of life" because it brings cells essentials such as (11) _____ , (12)

_____ from food, and secretions. It also takes away the (13) _____ produced by

metabolism. (14) _____ would be impossible were it not for our circulating blood.

        As the heart's pumping keeps pressure on blood flowing through the cardiovascular system, some water

and proteins are forced out of the vast network of capillaries and become part of the (15) _____ fluid. A

network called the (16) _____ system picks up excess interstitial fluid and reclaimable solutes, and

returns them to the cardiovascular system.

Labeling

In the following illustration, color in red all vessels that carry oxygen-rich blood (including all parts indicated by
"aorta" or "artery", except for the pulmonary arteries that carry oxygen-poor blood from the heart to the lungs).
Then fill in all the blanks on the right side of the diagram. Next, color in blue all vessels that carry oxygen-poor blood
(including all parts indicated by "vena cava" or "vein", except for the pulmonary veins that return oxygen-rich blood
to the heart). Then fill in all the blanks on the left side of the diagram. [p.122]

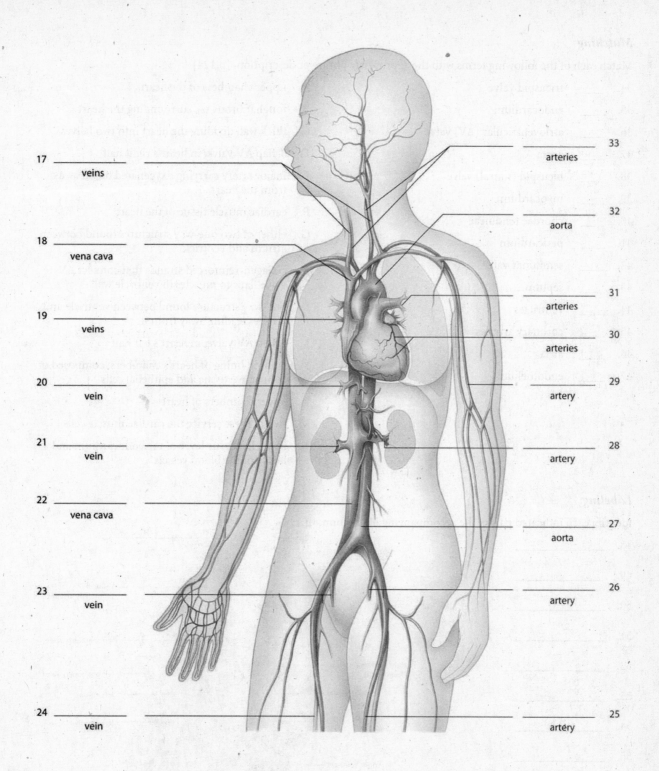

17 _____   33 _____
   veins   arteries

                                        32 _____
                                        aorta

18 _____   31 _____
   vena cava   arteries

19 _____   30 _____
   veins   arteries

20 _____   29 _____
   vein   artery

21 _____   28 _____
   vein   artery

22 _____   27 _____
   vena cava   aorta

                                        26 _____
23 _____   artery
   vein

24 _____   25 _____
   vein   artery

Circulation: The Heart and Blood Vessels   **107**

Matching

Match each of the following terms with the appropriate phrase or description. [p.124]

34. _____ tricuspid valve
35. _____ endocardium
36. _____ atrioventricular (AV) valve
37. _____ aorta
38. _____ bicuspid (mitral) valve
39. _____ myocardium
40. _____ chordae tendineae
41. _____ pericardium
42. _____ semilunar valve
43. _____ septum
44. _____ ventricles
45. _____ coronary arteries
46. _____ atria
47. _____ endothelium

A. upper chambers of the heart
B. tough, fibrous sac surrounding the heart
C. thick wall dividing the heart into two halves
D. 3-flap AV valve; in heart's right half
E. major artery carrying oxygenated blood away from the heart
F. cardiac muscle tissue of the heart
G. either of two one-way structures found between atrium and ventricle
H. collagen-reinforced strands that connect AV valve flaps to muscles in ventricle wall
I. one-way structure found between ventricle and arteries leading away from it
J. 2-flap AV valve; in heart's left half
K. smooth lining of heart's chambers; composed of connective tissue and epithelial cells
L. lower chambers of heart
M. two of these service the cardiac muscle cells
N. epithelial tissue layer of the endocardium and lining of the blood vessels

Labeling

Identify each indicated part of the accompanying illustration. [p.124]

48. _____
49. _____
50. _____
51. _____
52. _____
53. _____
54. _____
55. _____
56. _____

Sequencing

57. Show the correct sequence of the path of blood flow through the heart by numbering the events 1-8 in the order that they occur. The first event (1) is given. [pp.124-125]

__1__   Relaxed atria fill with blood.

_____   Ventricles contract and AV valves close ("lub").

_____   Atrioventricular (AV) valves open.

_____   Blood enters aorta and pulmonary artery.

_____   Ventricles relax and SL valves close ("dup").

_____   Blood flows into ventricles.

_____   Semilunar (SL) valves open.

_____   Atria contract.

Sequencing

Show the correct sequence of the path of blood flow through the pulmonary and systemic circuits by placing the letter of the part of the heart that receives blood from body tissues in the blank beside number 58, then continuing the sequence until you are back at the starting point. [pp.126-127]

58. _____   A. lungs

59. _____   B. aorta

60. _____   C. main pulmonary artery

61. _____   D. right atrium (use twice)

62. _____   E. left atrium

63. _____   F. right ventricle

64. _____   G. left ventricle

65. _____   H. pulmonary veins

66. _____   I. torso (systemic circulation)

67. _____   J. right and left pulmonary arteries

68. _____

Dichotomous Choice

Circle the correct answer of the two choices given between parentheses in each statement. [p.126]

Circulation: The Heart and Blood Vessels   **109**

69. Events 58-64 are parts of the (systemic / pulmonary) circuit.

70. Events 64-68 are parts of the (systemic / pulmonary) circuit.

Fill-in-the-Blanks [p.126-127]

As the aorta descends into the torso, major (71) _____ branch off it, funneling blood to (72)

_____ and tissues. In both the pulmonary and systemic circuits, blood travels through (73)

_____ , (74) _____ , (75) _____ , and (76) _____ , and finally

returns to the heart in (77) _____ . Blood from the head, arms, and chest arrives through the (78)

_____ , and the (79) _____ collects blood from the lower body. The actual exchange of

substances between blood and tissues occurs in (80) _____ .

After a meal, blood passing through capillary beds in the GI tract detours through the (81)

_____ to the liver. The liver removes (82) _____ and processes absorbed substances.

Blood leaves the liver's capillary bed through a (83) _____ vein. The liver receives (84)

_____ blood via the hepatic artery.

Dichotomous Choice

Circle the correct answer of the two choices given between parentheses in each statement. [p.128]

85. A signal to contract spreads rapidly through the heart because of gap junctions called (Purkinje fibers / intercalated discs) between abutting cells

86. The (cardiac conduction / intercalated disc) system produces the electrical impulses that stimulate contraction of the heart.

87. Contraction follows a wave of excitation that begins at a mass of cells in the upper wall of the right atrium called the (atrioventricular / sinoatrial) node.

88. After causing both atria to contract, the wave of excitation reaches the (atrioventricular / sinoatrial) node in the septum between the two atria where it slows down momentarily to give the atria time to finish contracting.

89. The ventricles then contract when the wave continues through conducting bundles that extend from the AV node and make contact with the muscle cells of each ventricle by means of conducting cells called (Purkinje fibers / sinoatrial nodes).

90. An artificial pacemaker is implanted when the cardiac pacemaker or (AV / SA) node malfunctions.

91. The (cardiovascular / nervous) system can adjust the rate and strength of cardiac muscle contraction.

7.5. BLOOD PRESSURE [p.129]

7.6. STRUCTURE AND FUNCTIONS OF BLOOD VESSELS [pp.130-131]

7.7. CAPILLARIES: WHERE SUBSTANCES MOVE BETWEEN BLOOD AND TISSUES [pp.132-133]

7.8. CARDIOVASCULAR DISEASES AND DISORDERS [pp.134-135]

7.9. INFECTIONS, CANCER, AND HEART DEFECTS [p. 136]

7.10. THE CARDIOVASCULAR SYSTEM AND BLOOD IN HOMEOSTASIS [p.137]

Boldfaced Terms

blood pressure _____

_____

hypertension _____

_____

pulse _____

_____

capillary beds _____

_____

vasodilation _____

_____

vasoconstriction_____

_____

baroreceptor reflex_____

_____

carotid arteries_____

_____

precapillary sphincter _____

_____

atherosclerosis _____

_____

plaque _____

_____

aneurysm _____

_____

heart attack _____

_____

arrhythmias _____

_____

atrial fibrillation (AFib)_____

_____

ventricular fibrillation_____

_____

rheumatic fever _____

_____

endocarditis _____

_____

Lyme disease _____

_____

myocarditis _____

_____

Fill-in-the-Blanks [p.129]

(1) _____ is the fluid pressure that blood exerts against vessel walls. Blood pressure is at its

highest in the (2) _____ , then it drops along the (3) _____ circuit. Consider an adult with

a blood pressure of 120/80; 120 is the (4) _____ pressure, which is the peak of pressure in the (5)

_____ when the heart's left ventricle pushes blood into it. The number 80 is the (6) _____

pressure, which is the lowest blood pressure in the (7) _____ when the heart is relaxed. Elevated blood

pressure, or (8) _____ , is associated with atherosclerosis and kidney disease. Low blood pressure is

called (9) _____ . A drastic drop in blood pressure, such as would occur from a large blood loss, is a sign

of (10) _____ .

Choice

For questions 11–24, choose from the following: [pp. 130-131]

    a. arteries   b. arterioles   c. capillaries   d. venules   e. veins

11. ____ Those near the body surface provide a "pulse".
12. ____ These merge into venules.
13. ____ Their walls have rings of smooth muscle over a single layer of elastic fibers.
14. ____ These serve as diffusion zones for exchanges between blood and tissue fluid.

15. ____ Their bulging elastic walls keep blood flowing on through the system.

16. ____ These offer more resistance to blood flow than other vessels do.

17. ____ These present less total resistance to flow than do the arterioles leading into them; the total drop in blood pressure is more gradual in this region.

18. ____ These serve as blood volume reservoirs.

19. ____ These allow diffusion of *some* solutes.

20. ____ These have valves that prevent backflow.

21. ____ These branch into arterioles.

22. ____ Weak valves in these leads to pooled blood and a varicose condition.

23. ____ Contractions of smooth muscle in their thin walls occur when blood must circulate faster.

24. ____ These merge into veins.

Labeling

Identify each vessel and the parts indicated by asterisks on the illustration below. [p.130]

25. _____; * _____ _____

26. _____; * _____ _____

27. _____; * _____

28. _____; * _____ _____

29. _____; * _____

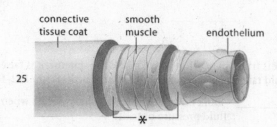

25

connective tissue coat   smooth muscle   endothelium

*

26

*   endothelium

27

*

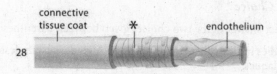

28

connective tissue coat   *   endothelium

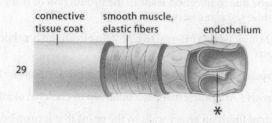

29

connective tissue coat   smooth muscle, elastic fibers   endothelium

*

Circulation: The Heart and Blood Vessels   113

Matching

Match the following terms with the correct definition. [p.131]

30. _____ vasodilation

31. _____ vasoconstriction

32. _____ baroreceptors

33. _____ brain centers

A. Pressure receptors in the carotid arteries, the aorta, and elsewhere that monitor changes in mean arterial pressure

B. Monitor resting blood pressure and alters the rate and strength of heartbeat

C. A decrease in blood vessel diameter brought about by brain centers when an abnormal decrease in blood pressure is detected

D. Enlargement of a blood vessel brought about by brain centers when an abnormal increase in blood pressure is detected

True/False

If the statement is true, write a T in the blank. If the statement is false, make it correct by writing the word(s) in the blank that should take the place of the underlined word(s). [pp.132-133]

34. _____ "Bulk flow" out of the capillary occurs when blood pressure inside a capillary is <u>lower</u> than fluid pressure outside.

35. _____ Because capillaries lie very close to nearly all your living cells, most solutes move from bloodstream to cells by <u>diffusion</u>.

36. _____ Water soluble substances can pass through <u>pores</u> in the capillary walls.

37. _____ The excess fluid that moves from capillaries into surrounding tissues is returned to the blood by the <u>cardiovascular</u> system.

38. _____ The precapillary sphincter regulates blood flow from <u>venules</u> into capillaries.

39. _____ When cells need more oxygen due to increased activity, precapillary sphincters <u>relax</u>.

Dichotomous Choice

Circle the correct answer of the two choices given between parentheses in each statement.[pp.134-135]

40. Recent studies indicate that (inflammation / high density cholesterol) is a trigger for the formation of artery-blocking plaques.

41. Inflamed tissue due to infection leads to the production of (homocysteine / C-reactive protein) that is a risk factor for cardiovascular disease.

42. Too much (homocysteine / C-reactive protein) from the breakdown of proteins causes damage that can lead to atherosclerosis.

43. (Atherosclerosis / Arteriosclerosis) is a thickening and stiffening of the arteries.

44. (Atherosclerosis / Arteriosclerosis) is a build-up of cholesterol and other lipids on artery walls.

45. The build-up of lipids on artery walls to the point that it protrudes into the artery is called a(n) (atherosclerotic / lipoprotein) plaque.

46. A blood clot floating in the bloodstream is called a(an) (thrombus/ embolus).

47. Proteins that pick up cholesterol in the blood and take it to the liver to be processed are (LDLs / HDLs).

48. Proteins that bind cholesterol and carry it to body cells are called (LDLs / HDLs).

49. Coronary bypass surgery involves stitching a large vessel taken from the chest to the aorta and past the point of damage in the (coronary vein / coronary artery).

50. (Balloon / Laser) angioplasty uses laser beams to vaporize plaques.

51. (Balloon / Laser) angioplasty uses an inflated balloon to flatten a plaque against an artery wall.

52. A pouch-like weak spot that causes the wall of an artery to balloon outward is a(n) (embolism / aneurysm).

53. Pain behind the breastbone, numbness or pain down the left arm and nausea are warning signs of heart (attack / failure).

54. A myocardial infarction is the medical term for heart (attack / failure).

55. A recording of the electrical activity of the cardiac cycle is an (MRI / ECG).

56. Bradycardia, tachycardia and ventricular fibrillation are types of (heart disease / arrhythmias).

57. A below average resting heart rate is called (bradycardia / tachycardia).

58. (Bradycardia / Tachycardia) occurs during exercise or stressful situations.

59. Dangerous haphazard contractions in parts of the ventricles are called ventricular (fibrillation / defibrillation).

Short Answer

60. List three strategies for improving your chances of avoiding serious cardiovascular disease. [p. 135] _____

_____

_____

_____

Fill-in-the-Blanks [p.136]

Untreated "Strep" infections may lead to (61) _____ . The body produces antibodies that

attack the invading bacteria and may also mistakenly attack and damage (62) _____ . This is an example

of an (63) _____ disorder.

Microbes may also enter the bloodstream during dental surgery or on a contaminated IV needles and attack

(64)_____ directly causing a condition called (65) _____ , ("inside the heart").

The tick borne bacterium, *Borrellia burgdorferi*, causes (66) _____ . A (67) _____

is the first bodily response and later joints may become inflamed, and so may the (68) _____ . Heart

inflammation is called (69) _____ . Measles caused by the rubella virus can also damage the (70)

_____ as can alcohol abuse and recreational drugs.

(71) _____ almost never starts in the heart muscle or blood vessels, but cancer that begins

elsewhere in the body may spread to the (72) _____ . Cancer treatment such as radiation or

chemotherapy may damage the heart or (73) _____ .

(74) _____ are infants born with a hole in some part of the heart wall, so that the heart doesn't

pump blood efficiently.

Circulation: The Heart and Blood Vessels   115

Short Answer

75. Explain how the cardiovascular system and blood contribute to homeostasis. [p.137] _____

_____

_____

_____

SELF-QUIZ

___ 1. The layer of cardiac muscle is the _____.
[p.124]

    a. pericardium
    b. endothelium
    c. endocardium
    d. myocardium
    e. myothelium

___ 2. _____ valves may be either bicuspid or tricuspid. [p.124]

    a. Sinoatrial
    b. Semilunar
    c. Atrioventricular
    d. Lymphatic
    e. Venous

___ 3. During systole, _____. [p.124]

    a. the aorta contracts
    b. both ventricles contract
    c. the entire heart relaxes
    d. only the left atrium and left ventricle contract
    e. only the right atrium and right ventricle contract

___ 4. The "lub-dup" sound of the heart comes from _____ [p.125]

    a. the SA valves closing followed by the AV valves closing
    b. the AV valves closing followed by the semilunar valves closing
    c. the semilunar valves closing followed by the SA valves closing
    d. the contraction of the atria followed by the contraction of the ventricles
    e. electrical impulses from the SA node

___ 5. Begin with a red blood cell located in the superior vena cava and travel with it in proper sequence as it goes through the following structures. Which will be last in sequence? [pp.126-127]

    a. aorta
    b. left atrium
    c. pulmonary artery
    d. right atrium
    e. right ventricle

___ 6. Arterioles and venules are connected by _____. [pp.130-131]

    a. thoroughfare channels
    b. interstitial passageways
    c. hepatic portals
    d. true capillaries
    e. renal portals

___ 7. The pacemaker of the human heart is the _____. [p.128]

    a. sinoatrial node
    b. semilunar valve
    c. inferior vena cava
    d. superior vena cava
    e. atrioventricular node

___ 8. _____ are blood reservoirs in which resistance to flow is low. [p.131]

    a. Arteries
    b. Arterioles
    c. Capillaries
    d. Venules
    e. Veins

___ 9. Which of these is NOT a risk factor for cardiovascular disease? [p.135]

    a. inflammation
    b. high cholesterol levels
    c. hypotension
    d. smoking
    e. obesity

___ 10. Deoxygenated blood is pumped from the _____ to the lungs. Oxygenated blood returns to the _____ of the heart. [126]

    a. right ventricle, right atrium
    b. left atrium, left ventricle
    c. right atrium, left ventricle
    d. right ventricle, left atrium
    e. left ventricle, right atrium

CHAPTER OBJECTIVES/REVIEW QUESTIONS

1. Name and briefly describe the two main elements of the cardiovascular system. [p.123]

2. In what way is the circulatory system linked with the lymphatic system? [p.123]

3. Describe the structure of the heart, including linings, chambers, and valves. [p.124]

4. The contraction phase of the cardiac cycle is called _____ [p.124] and the relaxation phase is called _____ [p.124].

5. Trace the path of blood in the human body. Begin with the aorta and name all major components of the circulatory system through which all blood passes before it returns to the aorta. [pp.126-127]

6. The _____ [p.126] circuit receives blood from the body tissues and circulates it through the lungs for gas exchange; the _____ [p.126] circuit transports blood to and from tissues.

7. Describe the role of capillary beds in the cardiovascular system. Where are these structures found in the body? [pp.126, 130-33]

8. Explain what causes the chambers of the heart to contract. [p.128]

9. Explain a blood pressure such as 118/76 in terms of systolic and diastolic pressure. Is this a healthy blood pressure? What medical terms refer to elevated blood pressure and low blood pressure? Which is more dangerous and why? [p.129]

10. Describe how the structures of arteries, arterioles, and capillaries differ; describe how the structures of venules and veins differ. [pp.130-131]

11. What types of things account for heart attacks in people with no obvious risk factors? [p.134]

12. Describe the process leading to atherosclerosis. What are the consequences of this condition? [p.134]

13. State the significance of high- and low-density lipoproteins to cardiovascular disorders. [p.134]

INTEGRATING AND APPLYING KEY CONCEPTS

1. Most heart attacks are due to blockages in coronary circulation. Why do they often result in lasting damage?

2. Consider what might be the result of faulty valves in the veins or the heart.

8

BLOOD

INTRODUCTION

Blood is not only the body's main transport mechanism, it also has a major function in maintaining homeostasis. This chapter examines the many substances contained in blood and details the various functions of the plasma, the red blood cells, white blood cells and platelets in the maintenance of homeostasis. Additional topics include hormonal control of blood cell production, blood typing, blood clotting and blood disorders.

FOCAL POINTS

- Figure 8.1 [p.142] summarizes the components of blood, their proportions and functions.
- Figure 8.2 animated [p.143] illustrates the stem cells of the bone marrow and the cells they produce.
- Figure 8.3 animated [p.144] highlights the structure of the hemoglobin molecule.
- Figure 8.4 [p.145] demonstrates the hormonal feedback mechanism responsible for erythrocyte production.
- Table 8.1 animated [p.146] provides a summary of antigens and antibodies in ABO blood types.
- Figures 8,5 [p.147] and 8.6 animated [p.148] illustrate the antigen-antibody response, and the importance of blood typing, both with ABO blood types and Rh factor. In addition, Fig. 8.6 demonstrates the dangers of Rh incompatibility in a pregnant woman.
- Table 8.2 [p.149] lists some of the tests that can be used to reveal imbalances in the body's homeostasis.
- Figures 8.7 [p.150] and 8.8 [p.151] summarize the steps involved in blood clotting and wound healing.

CHAPTER INTRODUCTION [p.141]

8.1. BLOOD: PLASMA, BLOOD CELLS AND PLATELETS [pp.142–143]

Boldfaced Terms

blood _____

_____

plasma _____

_____

erythrocytes _____

_____

red blood cells _____

_____

leukocytes _____

_____

white blood cells _____

_____

platelet _____

_____

Fill-in-the-Blanks [pp.142-143]

For the average-sized female adult, blood volume is generally about (1) _____ liters, which is 6 to 8 percent of body weight. About 55% of whole blood is (2) _____ , which is mostly (3) _____ and serves as a transport medium for blood cells, platelets, and other substances. Two-thirds of all plasma proteins are molecules of (4) _____ , which plays an important role in the osmotic movement of (5) _____ in and out of the blood. Other plasma (6) _____ include hormones, immune system proteins and proteins involved in blood clotting. Ions in the plasma help maintain the (7) _____ and (8) _____ of extracellular fluid .

(9) _____ cells or erythrocytes make up about 45 percent of whole blood. Each is a biconcave disk carrying the iron-containing protein, (10) _____ . These cells transport oxygen used in aerobic respiration as well as some carbon dioxide wastes. Red blood cells are derived from unspecialized (11) _____ located in the red bone marrow.

Leukocytes, or (12) _____ cells, make up a tiny fraction of whole blood, but have vital functions in day-to-day housekeeping and defense. They scavenge dead or worn-out cells, as well as any material identified as (13) _____ to the body. Others target or destroy (14) _____ like bacteria and viruses. They circulate in the blood, but most squeeze out of blood vessels and do their work after they enter (15) _____ . All white blood cells develop from stem cells located in the (16) _____ .

One group of leukocytes called (17) _____ contain various granules in the cytoplasm. This group includes (18) _____ , (19) _____ and (20) _____ . Leukocytes called (21) _____ have no visible granules in the cytoplasm. The first of two types are called (22) _____ , which differentiate into macrophages that engulf invaders and cellular debris. The second type, (23) _____ (B cells and T cells), carry out specific immune responses.

Some stem cells in bone marrow develop into "giant" cells called (24) _____ that shed fragments of cytoplasm that become enclosed in a bit of plasma membrane. The fragments are called (25) _____ . They release substances that initiate blood (26) _____ .

Dichotomous Choice

Circle the correct answer from the pair of choices given for each statement. [pp.142-143]

27. The cell count of (erythrocytes / leukocytes) increases when the body is fighting an infection.

28. (Platelets / plasma proteins) are made by shedding cellular fragments from a megakaryocyte.

29. The most numerous type of leukocyte is the (basophil / neutrophil).

30. Plasma proteins constitute approximately (8% / 15%) of the plasma volume.

31. The blood cells that operate during immune responses are (neutrophils / lymphocytes).

Identification

When blood is spun in a centrifuge, the components separate into regions as shown in the following diagram. Select the letter of the region where you will find the following blood components. [p.142]

32. _____ erythrocytes

33. _____ leukocytes

34. _____ plasma

8.2. HOW BLOOD TRANSPORTS OXYGEN [p.144]

8.3. MAKING NEW RED BLOOD CELLS [p.145]

Boldfaced Terms

hemoglobin _____

_____

oxyhemoglobin _____

_____

Complete Blood Count (CBC) _____

120   Chapter Eight

Interpreting Diagrams

Study each diagram below in order to answer the questions that follow. [p.144]

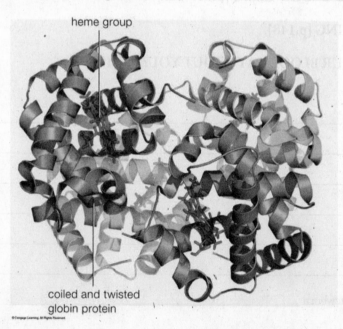

LUNGS

more O₂
cooler
less acidic

$Hb + O_2 \longrightarrow HbO_2$

TISSUES

$HbO_2 \longrightarrow Hb + O_2$

less O₂
warmer
more acidic

1. What is "Hb" in the diagram?_____

2. What is HbO₂? _____

3. Under what conditions (oxygen, temperature, pH) does oxygen tend to bind to hemoglobin?_____

_____

4. Where in the body do the conditions in question #3 exist? _____

5. Under what conditions (oxygen, temperature, pH) does hemoglobin tend to give up oxygen?_____

_____

6. Where in the body do the conditions in question #5 exist? _____

heme group

coiled and twisted
globin protein

© Cengage Learning. All Rights Reserved.

7. What is the molecule diagrammed above? _____

8. In what type of blood cells are molecules like this found? _____

9. How many polypeptide chains make up the globin component of this molecule? _____

10. What inorganic element is found at the center of each heme group? _____

11. What is the function of the molecule shown above? _____

12. What is this molecule called when it binds oxygen? _____

Matching

Select the description that best matches the term given. [p.145]

13. ____ erythropoietin

14. ____ bilirubin

15. ____ bone marrow

16. ____ 120 days

17. ____ stem cell

18. ____ complete blood count

19. ____ "blood doping"

20. ____ spleen

A. where red blood cells are produced

B. average lifespan of red blood cell

C. cells in marrow that give rise to red blood cells

D. average number of red blood cells, white blood cells and platelets per microliter

E. where macrophages remove and recycle old or damaged red blood cells

F. hormone made in kidneys that stimulates production of red blood cells

G. orangish remnant of heme group removed from bloodstream by liver

H. injection of stored blood of an athlete

8.4. DIFFERENT RED BLOOD CELLS [p.146-147]

8.5. Rh BLOOD TYPING [p.148]

8.6. WHAT CAN YOUR BLOOD SAY ABOUT YOU? [p.149]

Boldfaced Terms

ABO blood typing _____

_____

agglutination _____

_____

Rh blood typing _____

_____

hemolytic disease of the newborn _____

_____

Matching

Choose the most appropriate description for each term. [pp.146-148]

1. ____ antibodies
2. ____ antigen
3. ____ ABO typing
4. ____ Rh factor
5. ____ hemolytic disease of the newborn
6. ____ cross-matching
7. ____ agglutination
8. ____ Type AB
9. ____ Type O
10. ____ Types A and B

A. A person whose blood type is positive carries this marker

B. Blood analysis based on presence or absence of glycoproteins A and B

C. Theoretical "universal recipient"

D. Theoretical "universal donor" group

E. Blood types that carry only one ABO blood group marker

F. May occur in offspring produced by an Rh– woman and an Rh+ man

G. Method used to insure compatibility of blood marker forms

H. Clumping when antibodies attack foreign cells

I. Immune system proteins that recognize and organize an attack on most foreign entities

J. "Non-self" protein markers that prompt a defensive attack by antibodies

Completion

Fill in the missing information on ABO blood types in the table below. [p.146]

Blood Type	Antigens on Plasma Membranes of RBCs	Antibodies in Blood
A	11.	15.
B	12.	16.
AB	13.	17.
O	14.	18.

Dichotomous Choice

Choose the term in parentheses which best completes the sentence. [p.146-148]

19. Rhogam is used to treat an (Rh⁺ / Rh⁻) woman after the birth of her first child.

20. An autologous transfusion consists of blood donated by (a close relative / the person receiving the transfusion).

21. (Type AB/ type O) blood is the rarest type among all populations in the United States.

Matching

Match the blood test with its diagnostic function.
[p.149]

22. ____ BUN

23. ____ TSH

24. ____ A1C

25. ____ CBC

26. ____ lipid profile

A. anemias, infection

B. kidney function

C. risk of cardiovascular disease

D. thyroid disorders

E. diabetes

8.7. HEMOSTASIS AND BLOOD CLOTTING [pp.150-151]

8.8. BLOOD DISORDERS [pp.152-153]

Boldfaced Terms

hemostasis _____

_____

thrombosis _____

_____

embolism _____

_____

stroke _____

_____

hemophilia _____

_____

anemias _____

_____

malaria _____

_____

thalassemia _____

_____

infectious mononucleosis _____

_____

leukemias _____

_____

chronic myelogenous leukemia _____

_____

septicemia _____

_____

toxemia _____

_____

Sequence

The following lettered stages of the "intrinsic" clotting mechanism occur in a specific sequence. Place the letter of the first event in number 1, and continue until the last event is placed in blank number 6. [p.150]

1. ____

2. ____

3. ____

4. ____

5. ____

6. ____

A. Fibrinogen proteins stick together, forming an insoluble net of fibrin

B. A blood vessel ruptures

C. Platelets aggregate and form a temporary plug.

D. Activated Factor X triggers reactions that form thrombin

E. Smooth muscle in a damaged blood vessel contracts, constricting the vessel and slowing blood flow

F. Blood cells and platelets become entangled in the fibrin net, forming a clot

Matching

Select the lettered description that best matches the numbered term. [pp.150-151]

7. ____ Stroke

8. ____ Hemophilia

9. ____ Hemostasis

10. ____ Embolus

11. ____ Aspirin

12. ____ Thrombus

13. ____ Extrinsic Clotting Mechanism

14. ____ Intrinsic Clotting Mechanism

Blood   125

A. a genetic disorder in which blood fails to clot due to missing clotting factors

B. a blood clot blocks blood flow to a part of the brain and brain tissue dies

C. reactions leading to clotting are triggered by substances outside the bloodstream

D. reduces the clumping of platelets; used to prevent blood clots

E. clotting mechanism triggered by substances inside the blood itself

F. a homeostatic mechanism designed to stop bleeding

G. a traveling blood clot

H. a blood clot forms in an unbroken blood vessel and remains there

Matching

Select the lettered description that best matches the numbered term. [pp.152-153]

15. _____ iron-deficiency anemia

16. _____ aplastic anemia

17. _____ infectious mononucleosis

18. _____ sickle cell anemia

19. _____ hemolytic anemia

20. _____ MRSA

21. _____ pernicious anemia

22. _____ malaria

23. _____ septicemia

24. _____ leukemia

25. _____ toxemia

26. _____ thalassemia

A. Runaway multiplication of abnormal white blood cells due to cancer of the bone marrow

B. Premature destruction of red blood cells

C. Red blood cells burst after invasion by protozoan parasites transmitted by mosquitoes

D. Hemoglobin cannot be formed due to low iron supply

E. Strain of *Staphylococcus aureus* highly resistant to antibiotics

F. Destruction of red bone marrow by radiation or toxins

G. Red blood cells are fragile and made in low numbers due to abnormal hemoglobin

H. Metabolic poisons accumulate, often due to loss of kidney function

I. Caused by a deficiency of folic acid or vitamin B12

J. Overproduction of agranulocytes, caused by Epstein- Barr virus

K. Inherited disease in which shape of red blood cells can become distorted

L. Massive infection resulting from toxins released by bacteria in the bloodstream

SELF-QUIZ

___ 1. Most of the oxygen in human blood is transported by _____ . [p.143]
   a. plasma
   b. serum
   c. platelets
   d. hemoglobin
   e. leukocytes

___ 2. Agranulocytes called _____ operate against specific invaders in an immune response. [p.143]
   a. basophils
   b. eosinophils
   c. monocytes
   d. neutrophils
   e. lymphocytes

___ 3. Oxygen becomes bound to _____ in hemoglobin molecules. [p.144]
   a. iron in heme groups
   b. polypeptides
   c. plasma
   d. globin
   e. carbon dioxide molecules

___ 4. The process that stops bleeding is called _____. [p.150]
   a. thrombosis
   b. phagocytosis
   c. inflammation
   d. agglutination
   e. hemostasis

___ 5. Oxyhemoglobin is hemoglobin combined with _____. [p.144]
   a. carbon dioxide
   b. red blood cells
   c. plasma
   d. calcium
   e. oxygen

___ 6. Megakaryocytes shed millions of _____ into the blood that last about a week. [p.143]
   a. platelets
   b. red blood cells
   c. lymphocytes
   d. macrophages
   e. neutrophils

___ 7. Which of the following is caused by a genetic disorder? [p.152]
   a. malaria
   b. thalassemia
   c. pernicious anemia
   d. aplastic anemia
   e. leukemia

___ 8. What combination can lead to hemolytic disease of the newborn? [p.148]
   a. Rh^- mother, Rh^- fetus
   b. Rh^+ mother, Rh^+ fetus
   c. Rh^+ mother, Rh^- fetus
   d. Rh^- mother, Rh^+ fetus
   e. any of these

___ 9. Red blood cells and white blood cells develop from _____ cells in the bone marrow. [pp.142]
   a. stem
   b. hemolytic
   c. megakaryocyte
   d. oxyhemoglobin
   e. albumin

___ 10. About how many liters of blood does an adult woman have? [p.142]
   a. 1-2
   b. 4-5
   c. 10-12
   d. 15-20
   e. 24-45

___ 11. The proportion of plasma to cell content in the blood is ____? [p.142]
   a. 75% to 25%
   b. 35% to 65%
   c. 55% to 45%
   d. 65% to 35%
   e. 45% to 65%

___ 12. Oxyhemoglobin becomes hemoglobin by releasing oxygen to the tissues under which of the following conditions? [p.144]

    a. more oxygen; cooler; less acidic
    b. more oxygen; warmer; less acidic
    c. more oxygen; warmer; more acidic
    d. less oxygen; warmer; more acidic
    e. less oxygen; cooler; less acidic

___ 13. Erythropoeitin is made by the _____. [p.145]

    a. kidneys
    b. hypothalamus
    c. leukocytes
    d. erythrocytes
    e. bone marrow cells

___ 14. A person with type O blood has _____ in the blood plasma. [p.146]

    a. A antigen
    b. B antigen
    c. both A and B antigens
    d. both A and B antibodies
    e. neither A nor B antibodies

___ 15. Which disease is caused by the Epstein-Barr virus? [p.152]

    a. leukemia
    b. pernicious anemia
    c. hemophilia
    d. septicemia
    e. infectious mononucleosis

CHAPTER OBJECTIVES/REVIEW QUESTIONS

1. Name and describe the composition of the two major components of human blood, using percentages of volume. [p.142]

2. Name the most common plasma protein and describe its functions. [p.142]

3. State where erythrocytes, leukocytes, and platelets are produced. [p.143]

4. Contrast the two main types of leukocytes in terms of cell structure. Then, name and state the general functions of the three types of granulocytes and the two types of agranulocytes. [p.143]

5. Describe the process of conversion of hemoglobin to oxyhemoglobin then back to hemoglobin. [p.144]

6. _____ is a hormone produced by the kidneys to stimulate certain stem cells to produce red blood cells. [p.145]

7. Describe what happens to red blood cells when they are old or damaged. [p.145]

8. List the ABO and Rh protein markers found on the red blood cells of all eight major blood types (A⁺, B⁺, AB⁺, O⁺, A⁻, B⁻ AB⁻ ,O⁻. [pp.146-148]

9. List in sequence the chemical events that occur in the formation of a blood clot. [p.150]

10. When red blood cells contain a less-than-normal amount of hemoglobin, it is termed _____. [p.152]

11. Describe disorders of the blood related to nutrient deficiencies, heredity, pathogens and cancer. [pp.152-153]

12. Name some of the tests that can be performed on blood. What do their results signify? [p.149]

INTEGRATING AND APPLYING KEY CONCEPTS

1. Environmental toxins cause many types of disease. Consider the effects on the body of a toxin that affects the bone marrow cells.

2. What benefits might "blood doping" confer on an athlete? Why do you think the practice has been banned by many athletic associations and commissions?

9

IMMUNITY AND DISEASE

INTRODUCTION

Immunity is an amazing, albeit complex, topic. Before starting to read the chapter, flip through the sections to gain an idea of how the material is organized. Conquer the first and second lines of immune defenses first. Then approach the most versatile and long-lasting of our immune responses, the adaptive immune system. Be aware that these responses occur in our bodies regularly.

FOCAL POINTS

- Figure 9.2 [p.157] diagrams the lymphatic system. Consider what you know about the cardiovascular system from Chapter 7 as you observe the locations of lymph nodes and vessels.
- Figure 9.7 animated [p.161] illustrates the process of inflammation.
- Figure 9.8 animated [p.162] compares primary and secondary immune responses.
- Figures 9.12 animated and 9.13 animated [pp. 165-166] provide illustrations of antibody-mediated and cell-mediated immune responses. Taking the time to work through the diagrams is strongly advised.
- Figure 9.21 animated [p.173] illustrates infection of a cell by the HIV virus.
- Table 9.5 [p.175] lists statistics of some major infectious diseases.

INTERACTIVE EXERCISES

CHAPTER INTRODUCTION [p.155]

9.1. OVERVIEW OF BODY DEFENSES [pp.156-157]

9.2. THE LYMPHATIC SYSTEM [pp.158-159]

Boldfaced Terms

antigen _____

_____

immunity _____

_____

innate immunity _____

_____

adaptive immunity _____

_____

lymphocytes _____

_____

immune system _____

_____

cytokines _____

_____

 neutrophils _____

_____

basophils _____

_____

mast cells _____

_____

macrophages _____

_____

eosinophils _____

_____

dendritic cells _____

_____

B cells _____

_____

T cells _____

_____

 lymphatic system _____

_____

lymph _____

_____

lymph vascular system _____

_____

lymph nodes_____

_____

spleen_____

_____

thymus _____

_____

Short Answer

1. List five examples of pathogens. [p.156] _____

_____

_____

_____

_____

2. How does the body identify these pathogens as non-self? [p.156] _____

_____

_____

_____

3. What term refers to the body's ability to combat the pathogens that it detects? [p.156] _____

_____

_____

_____

Choice

For questions 4-13, choose from the following possible answers. [p.156]

    a. body surface defense   b. innate immune system   c. adaptive immune system

4. _____   intact skin and linings of body cavities

5. _____   specialized lymphocytes and proteins to attack invading microbes

6. _____ response to a pathogen detected internally which begins within minutes

7. _____ preset (inborn) responses to invasion of pathogens

8. _____ second line of defense – internal, fast

9. _____ first line of defense - external

10. _____ third line of defense – internal, slow

11. _____ involves recognition of a specific pathogen

12. _____ general response to tissue damage

13. _____ leaves behind immune memory of a pathogen

Fill-in-the-Blanks [p.156]

The immune system relies on several types of chemicals to fight invaders. White blood cells secrete several types of chemicals called (14) _____. (15) _____ and (16) _____ contain chemicals in their cytoplasmic granules that promote inflammation. B cells make defensive proteins called (17) _____.

The immune response is stronger in (18) _____ than in (19) _____. Because of this, they are more likely to develop (20) _____ diseases like (21) _____, _____ or _____.

Matching

Match the type of white blood cell to its general function. [p.157]

22. _____ T cells

23. _____ macrophages

24. _____ dendritic cells

25. _____ neutrophils

26. _____ eosinophils

27. _____ basophils and mast cells

28. _____ B cells

A. Make defensive proteins called antibodies

B. Alert adaptive immune system of the presence of antigens

C. Release chemicals that cause inflammation

D. Some kill abnormal body cells while others activate B cells

E. Two-thirds of white blood cells

F. Target worms, fungi, and other large pathogens

G. Large phagocytes that arise from monocytes

Labeling

Identify the numbered parts of the accompanying illustration. [p.158]

29. _____

30. _____

31. _____

32. _____

33. _____

34. _____

35. _____

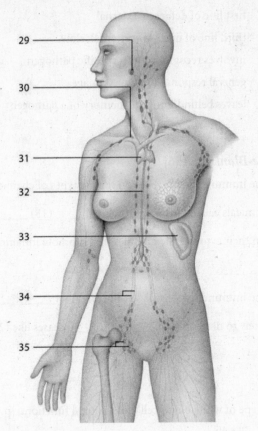

Short Answer

36. What is the source of tissue fluid? When is it referred to as lymph? [p.158] _____

_____

_____

_____

Choice

Indicate the function of the lymph vascular system to which each statement refers. Use the following choices. [p.159]

    a. drainage   b. delivery   c. disposal

37. _____ removes and carries foreign material and cellular debris
38. _____ collects water and solutes that have leaked out of capillaries
39. _____ picks up absorbed fats from small intestine and delivers them to bloodstream
40. _____ returns water and solutes to bloodstream
41. _____ transports fats to bloodstream
42. _____ delivers foreign substances to lymph nodes

Sequencing

43. Place the numbers 1 through 4 in the blanks to indicate the order of lymph flow. [p.159]

    _____   Lymph enters lymph vessels with valves to prevent backflow.

    _____   Fluid passes through "valves" into lymph capillaries at blood capillary beds.

    _____   Lymph returns to circulatory system at veins of lower neck.

    _____   Lymph moves into collecting ducts.

Choice

Indicate the lymphoid organ to which each statement refers. Use the following choices. [p.159]

    a. lymph node   b. spleen   c. thymus

44. _____ has red pulp as a reservoir of red blood cells and macrophages
45. _____ where T cells multiply and specialize against specific antigens
46. _____ has chambers containing white blood cells
47. _____ largest lymphoid organ
48. _____ site where, during an infection, lymphocytes destroy foreign agents
49. _____ has white pulp containing masses of lymphocytes

50. Name four places that lymphoid tissue is found, besides the lymph nodes, spleen and thymus. [p.159] _____

_____

_____

_____

_____

9.3. BARRIERS TO INFECTION [p.160]

9.4. INNATE IMMUNITY [pp.160-161]

9.5. OVERVIEW OF ADAPTIVE DEFENSES [pp.162–163]

Boldfaced Terms

lysozyme _____

_____

complement system _____

_____

membrane attack complexes _____

_____

inflammation _____

_____

histamine _____

_____

fever _____

_____

plasma cell _____

_____

effector cells _____

_____

136   Chapter Nine

memory cells _____

_____

antibodies _____

_____

antibody-mediated immunity _____

_____

helper T cells _____

_____

MHC markers _____

_____

cytotoxic T cells _____

_____

cell-mediated immunity _____

_____

antigen-presenting cell _____

_____

B cell receptors_____

_____

T cell receptors _____

_____

Short Answer

1. Explain how "friendly bacteria" on our skin and mucus membranes act as a defense against harmful species. What can happen if the populations of friendly bacteria are greatly reduced? [p.160]_____

_____

_____

_____

_____

Matching

Match the correct description with the following examples of surface defenses. [p.160]

2. _____ urine pH
3. _____ *Lactobacillus*
4. _____ lysozyme
5. _____ diarrhea

A. Blocks entry of many pathogens into urinary tract

B. Mild form rids lower GI tract of pathogens

C. Enzyme in mucus and tears that attacks bacteria

D. Maintains a low vaginal pH that many pathogens cannot tolerate

Fill-in-the-Blanks [p.160]

Phagocytosis is an important component of (6) _____ immunity. When a pathogen has entered the body, (7) _____ in tissue fluid are usually first on the scene. They engulf and destroy virtually (8) _____ other than healthy body cells. If they detect an antigen, they release (9) _____ which attract more (10) _____ as well as other white blood cells.

Also contributing to the phagocytosis response is the group of proteins called (11) _____ . These chemically detect the presence of (12) _____ and activate more complement proteins. Activated complement molecules attract (13) _____ to damaged tissues. They blanket pathogens with a (14) _____ . An invader that is coated with complement sticks to a (15) _____ which ingests it. Some complement proteins form (16) _____ that kill pathogens by making holes in their cell membranes.

Dichotomous Choice

Circle the correct answer of the two choices given between parentheses in each statement. [p.161]

17. Activated complement and (antigens / cytokines) secreted by macrophages trigger acute inflammation.

18. (Macrophages / Mast cells) respond to complement or an antigen by releasing histamines and cytokines.

19. Histamines cause arterioles to dilate so that (less / more) blood flows to an area, causing redness.

20. Histamines make capillaries leaky, resulting in swelling called (edema / necrosis).]

21. Fever develops when (antigens / interleukins) stimulate the brain to release (thermostatins / prostaglandins).

Short Answer

22. Explain how a fever can be beneficial. [p.161] _____

_____

_____

_____

Sequencing

Sequence the events listed and write their letters in the appropriate blanks to show the order in which they occur. [p.162]

A. Repeated division of activated B and T cells

B. Specialization of B and T lymphocytes as memory cells for future attacks of the same invader

C. Specialization of B and T cells lymphocytes as effector cells to fight the invader

D. Recognition of "nonself" marker by B and T lymphocytes

23. _____

24. _____

25. _____   26. _____

Short Answer

27. Which line of defense are B and T cells a part of? List the four terms that describe this line of defense. [p.162] __

_____

_____

_____

_____

Complete the Table

Compare B and T cells by completing the table below. [p.162-163]

	B cells	Helper T cells	Cytotoxic T cells
site of maturation	28.	29.	30.
location after maturing	31.	32.	33.
type of immunity	34.	*helps with antibody and cell-mediated*	35.

True/False

If the statement is true, write a T in the blank. If the statement is false, make it correct by writing the word(s) in the blank that should take the place of the underlined word(s). [p.162-163]

36. _____ All body cells have MHC markers.

37. _____ MHC markers are recognized by receptors on B cells.

38. _____ An antigen will be invisible to T cells until processed by an antigen presenting cell.

39. _____ Macrophages, dendritic cells and B cells can all present antigens.

40. _____ An antigen-presenting cell sticks to something bearing an antigen.

41. _____ Enzymes of the antigen-presenting cell break the antigen into pieces that are joined with lysosomes.

42. _____ Antigen-MHC complexes are displayed at the cell's surface.

43. _____ Helper T cells release cytokines.

44. _____ Effector B cells are called plasma cells.

9.6. ANTIBODY-MEDIATED IMMUNITY: DEFENDING AGAINST THREATS OUTSIDE CELLS [pp.164-165]

9.7. CELL-MEDIATED RESPONSES: COMBATING THREATS INSIDE CELLS [pp.166–167]

Boldfaced Terms

immunoglobulins _____

_____

apoptosis _____

_____

NK (natural killer) cells _____

_____

_____

Fill-in-the-Blanks [p.164]

If a B cell is activated, the resulting plasma cells produce (1)_____. Each has a (2)

_____ shape and can bind a(an) (3) _____ near the tip of the two "arms". Its

characteristics are determined by (4) _____ in a manner insuring that no two B cells will have identical

antibodies. As antibodies form, they are embedded in the B cells (5) _____ so that the two (6)

_____ of each stick outward.

As the antibody-mediated immune response begins, (7) _____ proteins coat the invading bacterial cell. It is then carried in the lymph to a(an) (8) _____ where it encounters B cells. The B cells (9) _____ it which moves the antigen into the B cell. A(an) (10) _____ forms and is displayed on the B cell surface. Next, receptors of a (11) _____ T cell bind to the complex, and the cells exchange signals. The T cell releases cytokines that (12) _____ the B cell and spur it to (13) _____.
Some of the B cell's descendants become (14) _____ cells that release huge numbers of antibodies that bind to antigens and flag them for destruction by (15) _____ and complement proteins. Other B cell descendants become (16) _____ cells that are available to respond rapidly to the antigen if it attacks the body at a later time.

Plasma cells produce (17) _____ classes of antibodies, collectively called (18) _____.

Matching

Select the description that best matches the numbered term. [p.165]

19. _____ IgG

20. _____ IgA

21. _____ IgM

22. _____ IgE

23. _____ IgD

A. Binds to basophils and mast cells, prompting histamine release and inflammation; involved in allergic reactions

B. Ten binding sites, first antibody secreted in immune responses and first made by newborns

C. B cell receptor

D. Found in tears, saliva, breast milk, and mucus

E. Most efficient at turning on complement proteins, neutralizes many toxins; most common antibody in bloodstream

True/False

If the statement is true, write a T in the blank. If the statement is false, make it correct by writing the word(s) in the blank that should take the place of the underlined word(s). [pp.166-167]

24. _____ Antibody-mediated responses defend the body against pathogens that can enter cells as well as against abnormal body cells such as cancer cells.

25. _____ A cell-mediated response begins when an antibody-presenting cell presents an antigen to T cells.

26. _____ Helper T cells produce cytokines that stimulate NK (natural killer) cells.

27. _____ NK cells don't need to have an antigen presented to them.]

28. _____ NK cells don't kill body cells flagged with stress markers.

29. _____ Helper T cells release molecules that directly kill infected and abnormal body cells.

30. _____ Cytotoxic T cells cause apoptosis.

31. _____ Cytotoxic T cells will cause rejection of transplanted corneas and testicles.

32. Explain why recipients of organ transplants must receive drugs to suppress the immune system. [p.167] _____

_____

_____

_____

_____

9.8. APPLICATIONS OF IMMUNOLOGY [pp.168-169]

9.9. IMMUNE SYSTEM DISORDERS [pp.170-171]

9.10. HIV AND AIDS [pp.172-173]

9.11. PATTERNS OF INFECTIOUS DISEASE [pp.174-175]

Boldfaced Terms

vaccine _____

_____

active immunity _____

_____

passive immunity_____

_____

monoclonal antibodies_____

_____

immunotherapy_____

_____

interferons_____

_____

multiple sclerosis _____

_____

allergens _____

_____

allergy _____

_____

hay fever _____

_____

anaphylactic shock _____

_____

immunological tolerance _____

_____

autoimmunity _____

_____

rheumatoid arthritis (RA) _____

_____

type 1 diabetes _____

_____

systemic lupus erythematosus (SLE) _____

_____

immunodeficiency_____

_____

severe combined immune deficiency (SCID) _____

_____

nosocomial infection _____

_____

epidemic _____

_____

pandemic _____

_____

sporadic disease _____

_____

endemic disease _____

_____

virulence _____

_____

Matching

Select the definition that best fits the numbered term. [pp. 168-169]

1. ____ monoclonal antibodies

2. ____ gamma interferon

3. ____ active immunity

4. ____ passive immunity

5. ____ immunotoxin

6. ____ "plantibody"

7. ____ booster shot

8. ____ beta interferon

9. ____ transgenic

10. ____ vaccine

A. Vaccine is injected or taken orally

B. Elicits a secondary response that results in more effector and memory cells being made

C. Being used to treat a type of multiple sclerosis

D. Genetically engineered viruses used to make vaccines

E. Injections of purified antibody molecules; for people who are already infected with pathogens

F. Genetically engineered plant being used to make cost-effective and safe antibodies

G. B cells cloned from a single antibody-producing cell; used to target some types of cancer cells

H. Genetically engineered form is used to treat hepatitis C

I. Combination of a poison and a monoclonal antibody; binds to and shuts down cancer cells

J. Prepared substance containing an antigen; elicits a primary immune response

Fill-in-the-Blank [pp.170-172]

A(n) (11) _____ is an immune response to a normally harmless substance. Such a substance is

called a(n) (12) _____ . Some people are genetically (13) _____ to develop allergies. When

an allergic person is first exposed to certain antigens, IgE (14) _____ are secreted and bind to (15)

_____ cells. These secrete prostaglandins, (16) _____ , and other substances that cause

inflammation. They also cause the person's airways to (17) _____ . (18) _____ produces

stuffy sinuses, a drippy nose and sneezing.

Food allergies are skewed responses of the immune system in which a particular food is interpreted as an

(19) "_____" . A whole-body allergic response to an allergen is called (20) _____ Air

passageways constrict, severe edema occurs, and blood pressure plummets, which can lead to complete collapse of

the (21) _____ system. The emergency treatment is an injection of the hormone (22) _____.

(23) _____ are anti-inflammatory drugs that counteract the release of histamines by mast cells.

In a desensitization program, doses of allergens are administered so that a person's body produces more circulating

(24) _____ molecules and memory cells, thus blocking allergic inflammation.

A(n) (25) _____ response is a disorder in which the body mobilizes its forces against normal

body cells or proteins. (26) _____ is an example of this kind of disorder, in which skeletal joints are

chronically inflamed. Another example is type 1 (27) _____ in which the immune system destroys the

insulin-secreting cells of the pancreas. In the autoimmune disease (28) _____ the affected person

develops antibodies to her or his own DNA and other "self" components. Immunodeficiency refers to disorders in

which the body does not have enough functioning (29) _____ . Both T and B cells are in short supply in

(30) _____ (SCID), an inherited life-threatening disorder. Infection by the (31) _____

(HIV) causes AIDS – (32) _____ .

HIV is transmitted when (33) _____ of an infected person enter another person's tissues. HIV

infects cells such as (34) _____ , (35) _____ cells , and (36)_____ cells that

have a certain type of surface receptor. The body then becomes dangerously susceptible to opportunistic infections

and to some otherwise rare forms of (37) _____ .

True/False

If the statement is true, write a T in the blank. If the statement is false, make it correct by writing the word(s) in the black that should take the place of the underlined word(s). [pp.172-173]

38. _____ There are many ways to rid the body of the known forms of the virus, HIV-I and HIV-II.

39. _____ Diagnostic signs of AIDS include having an active immune system, a positive HIV test, and having an "indicator" disease.

40. _____ "Indicator" diseases include types of pneumonia, recurrent yeast infections, cancer, and drug-resistant tuberculosis.

41. _____ HIV is transmitted when body fluids, especially blood and semen of an infected person enter another person's tissues.

42. _____ The most common mode of HIV transmission is sex with an infected partner.

43. _____ HIV is effectively transmitted by food, air, water, casual contact, or insect bites.

44. _____ HIV is a retrovirus, which means that its primary genetic instructions are in the form of DNA.

45. _____ The enzyme, reverse transcriptase, uses RNA as a template for making DNA. Eventually the DNA is "rewritten" back into RNA and these RNA instructions are then translated into RNA, and these RNA instructions are then translated into <u>nucleic acids</u>.

46. _____ There are <u>many</u> ways to remove HIV genes that are inserted into someone's DNA.

47. _____ HIV mutates <u>slowly</u>, so it can rapidly develop resistance to drugs.

48. _____ At present, the preferred HIV treatment is a drug "cocktail" that often consists of a <u>protease inhibitor</u> and two anti-HIV drugs.

49. _____ At present, the only real option for halting the spread of HIV appears to be prevention.

Short Answer

50. List four ways infectious disease can be spread from person to person. [p.174] _____

_____

_____

_____

51. Define nosocomial infection and give reasons they are so common. [p.174] _____

_____

_____

_____

_____

Matching

Select the description that best matches the numbered term. [p.175]

52. ____ epidemic

53. ____ pandemic

54. ____ virulence

55. ____ sporadic disease

56. ____ endemic disease

A. measure of pathogen damage and speed of invasion

B. breaks out irregularly and affects relatively few people

C. occurs more or less continuously

D. a disease rate increase to a level above what is predicted based on experience

E. epidemics break out in several countries around the world in a given time

SELF-QUIZ

Multiple Choice

____ 1. All the body's white blood cells are derived from stem cells in the _____. [p.158]

    a. spleen
    b. liver
    c. thymus
    d. bone marrow
    e. thyroid

____ 2. What type of immunity is represented by preset immune responses with which we are born? [p.156]

    a. surface immunity
    b. complement immunity
    c. innate immunity
    d. 1st line immunity
    e. adaptive immunity

____ 3. What surface barrier enzyme is found in tears and mucus? [p.160]

    a. lysozyme
    b. cytokine
    c. mucin
    d. interferon
    e. interleukin

____ 4. Which body system collects tissue fluid, cleans it, and returns it to the circulatory system? [p.158]

    a. immune
    b. cardiovascular
    c. nervous
    d. lymphatic
    e. endocrine

____ 5. Antibody molecules are shaped like the letter _____. [p.164]

    a. *Y*
    b. *W*
    c. *Z*
    d. *H*
    e. *E*

____ 6. Which lymphoid organ contains red and white pulp? [p.159]

    a. thymus
    b. lymph node
    c. bone marrow
    d. tonsils
    e. spleen

____ 7. B and T lymphocytes that are held in reserve are called _____. [p.162]

    a. effector cells
    b. macrophages
    c. mast cells
    d. helper cells
    e. memory cells

____ 8. Cell-mediated immunity involves _____ that directly attack target cells. [p.163]

    a. helper T cells
    b. effector B cells
    c. cytotoxic T cells
    d. cytokines
    e. plasma cells

____ 9. When an allergic person is first exposed to certain antigens, agent, IgE antibodies cause _____. [p.170]

    a. inflammation
    b. clonal cells to be produced
    c. B cell division
    d. the immune response to be suppressed
    e. an autoimmune disorder to develop

____ 10. What type of immunization is a vaccine? [p.168]

    a. passive
    b. active
    c. monoclonal
    d. autoimmune
    e. transductive

Matching

Choose the most appropriate description for each term.

11. _____ autoimmunity[p.171]

12. _____ antibody [p.164]

13. _____ antigen [p.156]

14. _____ macrophage [p.157]

15. _____ memory cells [p.162]

16. _____ complement [pp.160]

17. _____ histamine [p.160]

18. _____ MHC marker [p.162]

19. _____ effector cells [p.162]

20. _____ T cell [p.162]

A. Begins its development in bone marrow, but matures in the thymus gland

B. Cells that confer immunity following an initial infection

C. A chemical that causes blood vessels to dilate and let plasma proteins leak through the vessel walls

D. Y-shaped immunoglobulin produced by a B cell

E. A nonself marker that triggers the formation of lymphocytes

F. Produced by activated B and T cells to destroy a specific invader

G. A group of proteins that participate in innate and adaptive immune responses

H. A disorder in which the body's immune system attacks its own cells and proteins

I. Proteins sticking out of the plasma membrane of body cells that identify "self"

J. "Big eater" that lives for months and phagocytizes foreign agents

CHAPTER OBJECTIVES/REVIEW QUESTIONS

1. List and differentiate between the three lines of immune defenses present in humans. [p.156]

2. List and differentiate between the main types of chemical weapons of immunity. [p.157]

3. What are the main functions of the lymphatic system? [p.158]

4. What are the components of the lymph vascular system? [p.159]

5. Name and summarize the functions of the three main lymphoid organs. [p.159]

6. Describe typical surface barriers that organisms such as humans present to invading organisms. [p.160]

7. List and discuss innate immune responses that serve to exclude microbes from the body. [pp.160-161]

8 Explain how the complement system is involved in destroying invaders and fanning inflammation. [p.160]

9. Describe what occurs in the course of an inflammation, and how these events stop the spread of a pathogen. [pp.160-161]

10. What role does fever play in innate immunity? [p.161]

11. List the four key features steps of adaptive immunity. [p.162]

12. Distinguish between a cell-mediated response and an antibody-mediated response. [p.162-163]

13. Distinguish between effector and memory cells. [p.162]

14. How do B and T cells interact with antigen-presenting cells? [p.163]

15. What are immunoglobulins? What accounts for their diversity? [pp.164-165]

16. How are cytotoxic T cells involved in apoptosis? [p.166]

17. How are cytotoxic T cells involved in the rejection of transplanted organs? [p.167]

18. Explain how passive and active immunizations work. [pp.168-169]

19. How are monoclonal antibodies made and used? [p.169]

20. What is immunotherapy, and what are its applications? [p.169]

21. Distinguish allergy from autoimmune disease. [pp.170–171]

22. Describe how AIDS specifically interferes with the human immune system. [p.172-173]

23. Explain how an understanding of infectious disease transmission can help in prevention. [p.174]

INTEGRATING AND APPLYING KEY CONCEPTS

1. Vaccines can be developed against many pathogens, but not all. Diseases like the common cold and AIDS cannot be prevented by vaccination. Others, like influenza, must be vaccinated against every year. Humans are not capable of developing lasting immunity to any of the pathogens causing these diseases. Offer a possible genetic explanation for this variation among pathogens.

2. While visiting an equatorial area of the world, you notice that some people appear as though fluid has accumulated in their lower legs and feet. Their lower extremities resemble those of elephants. You inquire about what is wrong and are told that the condition is caused by a parasite injected by the bite of a mosquito. Construct a testable hypothesis that would explain (1) why the fluid was not being returned to the torso as normal, and (2) how the parasite might prevent the return of fluid.

10

THE RESPIRATORY SYSTEM

INTRODUCTION

After surveying the components of the respiratory system, this chapter takes you through the process of breathing and how it is regulated. While reading through this process, you will quite likely find yourself consciously inhaling and exhaling, becoming aware of the pathway of oxygen to your lungs and into your circulatory system. As you do, consider all of the homeostatic mechanisms that regulate the rate and depth of breathing.

FOCAL POINTS

- Figure 10.1 animated [p.180] diagrams the components of the respiratory system. As you study this figure, trace the pathways of oxygen and carbon dioxide in and out of the body.
- Figure 10.9 animated [p.184] illustrates the muscle movements necessary to ventilate the lungs. Especially take into account the role of the diaphragm in this process.

INTERACTIVE EXERCISES

CHAPTER INTRODUCTION [p.179]

10.1. THE RESPIRATORY SYSTEM: BUILT FOR GAS EXCHANGE [pp.180-181]

Boldfaced Terms

respiratory system _____

_____

pharynx _____

_____

larynx _____

_____

epiglottis _____

_____

trachea _____

_____

bronchus _____

_____

vocal cords _____

_____

lungs _____

_____

diaphragm _____

_____

pleurae (singular: pleura) _____

_____

bronchioles _____

_____

alveolus (plural: alveoli) _____

_____

Fill-in-the-Blanks [p.179]

Tobacco smoke is (1) _____ to human tissues.   Not only is cigarette smoke a major risk factor

for (2)_____, but it is also linked to other cancers. Females who begin smoking in their teens are (3)

_____ % more likely to develop breast cancer than non-smokers. Smoking (4) _____ the risk of

heart disease. Other effects of smoking are increased (5) _____ , higher levels of (6) _____

and lower levels of (7) _____ .

Labeling-Matching

Identify each of the components of the human respiratory system in the accompanying illustration by entering the correct names in the numbered blanks provided. Complete the exercise by matching and entering the letter of the correct function of each component in the parentheses that follow most of the labels. [p.180]

8. _____ ( )

9. _____ ( )

10. _____ ( )

11. _____ ( )

12. _____ ( )

13. _____ ( )

14. _____ ( )

15. _____ ( )

16. _____ ( )

17. _____ ( )

18. _____ ( )

19. _____

20. _____

21. _____ ( )

22. _____

23. _____

A. Airway where sound is produced; closed off while swallowing

B. Muscle sheet between chest cavity and abdominal cavity with roles in breathing

C. Increasingly branched airways between two bronchi and alveoli

D. Closes off larynx during swallowing

E. Chamber in which air is warmed, moistened, and filtered, and in which sound resontes

F. Ribcage muscles with roles in breathing

G. Airway that connects larynx with two bronchi

H. Supplemental airway when breathing is labored

I. Thin-walled air sacs where gases are exchanged between lungs and pulmonary capillaries

J. Membranes that separate lungs from other organs; fluid-filled cavity between has roles in breathing

K. Airway that connects nasal cavity and mouth with larynx; enhances sounds; also connects with esophagus

L. Lobed, elastic organ of breathing that enhances gas exchange between the internal environment and the outside air

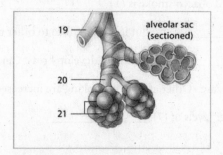

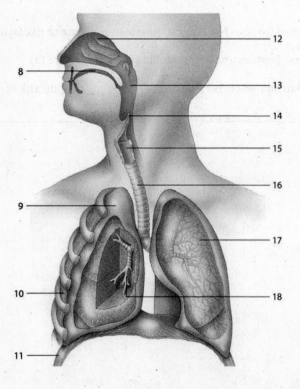

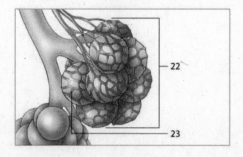

Matching

Choose the most appropriate description for each term. [pp.180-181]

24. _____ nose
25. _____ pleural membrane
26. _____ alveoli
27. _____ larynx
28. _____ vocal cords
29. _____ epiglottis
30. _____ bronchus
31. _____ lung capillaries
32. _____ intrapleural space
33. _____ intrapleural fluid
34. _____ bronchioles
35. _____ diaphragm
36. _____ glottis
37. _____ lung
38. _____ trachea

A. A sheet of muscle between the thoracic and abdominal cavities

B. Where air is filtered, warmed and moisturized before entering the respiratory system

C. Tiny blood vessels around alveoli that exchange oxygen and carbon dioxide with lungs

D. A gap between the vocal cords that is forced open with each exhalation

E. Where gas diffusion between lungs and lung capillaries takes place

F. Narrow airways with alveoli bulging from ends

G. Left one has two lobes, right one has three

H. Formed by a thin, double membrane of epithelium covering each lung

I. Covers glottis during swallowing to prevent choking

J. Consists of horizontal folds of mucus membrane near entrance to larynx

K. Formed of nine pieces of cartilage, one of which is the "Adam's apple"

L. "Windpipe" supported by rings of cartilage

M. Lubricating substance filling the intrapleural space

N. Branches that connect trachea to lungs; lined by cilia and cells that secrete mucus to trap bacteria and particles

O. A very narrow space between the two pleural membrane

Sequence

Arrange the following parts of the respiratory system in order in which air would enter them after leaving the nasal cavity. Place the letter of the first by the number 39, the next by the number 40, etc. [pp.180-181]

39. _____
40. _____
41. _____
42. _____
43. _____
44. _____
45. _____

A. alveolus
B. larynx
C. bronchus
D. pharynx
E. glottis
F. bronchiole
G. trachea

10.2. RESPIRATION = GAS EXCHANGE [pp.182-183]

10.3 BREATHING AT HIGH ALTITUDE AND UNDERWATER [p.183]

10.4. BREATHING: AIR IN, AIR OUT [pp.184-185]]

Boldfaced Terms

respiration _____

_____

respiratory surface _____

_____

respiratory cycle _____

_____

inspiration _____

_____

expiration_____

_____

pneumothorax _____

_____

tidal volume_____

_____

vital capacity _____

_____

Fill-in-the-Blanks]pp.182-183]

(1) _____ relies on the tendency of oxygen and carbon dioxide to diffuse down their

concentration gradients, or as we say in the case of gases, their (2) _____ gradients. The partial pressure

of oxygen at sea level is (3) _____ percent of 760 mm Hg (atmospheric pressure at sea level), which

equals 160 mm Hg. The partial pressure of (4) _____ is about 0.3 mm Hg. Large animals such as

humans must be capable of efficient gas exchange. Gases enter and leave the body by crossing a thin, moist (5)

_____ surface of epithelium. The surface must be moist because gases cannot diffuse across membranes

unless they are (6) _____ in fluid. The larger the surface area and the steeper the partial (7)

_____ , the faster diffusion occurs. In healthy human lungs, millions of thin-walled (8)

_____ provide a huge surface area for gas exchange. Gas exchange is boosted by the (9)

_____ in red blood cells. In the lung capillaries, each hemoglobin molecule binds with up to

(10) _____ oxygen molecules. When blood carries red blood cells into tissues where oxygen

concentration is low, hemoglobin (11) _____ oxygen. By carrying oxygen away from the respiratory

surface, hemoglobin helps maintain the required pressure (12) _____ that helps draw oxygen into the

lungs—and into blood in lung capillaries.

Dichotomous Choice

Circle one of two possible answers given between parentheses in each statement. [p.183]

13. Partial pressure of oxygen (decreases / increases) as altitude increases.

14. Higher than 2,400 meters, or about 8000 feet, brain respiratory centers compensate for oxygen deficiency by triggering faster and deeper breathing called (hypoxia / hyperventilation).

15. Raptures of the deep and decompression sickness result from the (decrease / increase) in pressure as a person descends into the water.

Analyzing diagrams

To better understand the mechanisms of respiration, study each indicated part of the following illustration. Then answer the accompanying questions. As you do the exercise, it aids understanding if you are conscious of the same parts of your own anatomy presently undergoing the respiratory cycle. [p.184]

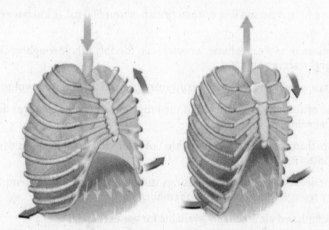

16. Which diagram, right or left, shows inspiration? _____

17. Which diagram, right or left, shows expiration? _____

18. Which muscles contract and relax to cause the rib cage movement indicated by the three larger arrows on each diagram? _____

19. The small arrows on each diagram show the contraction and relaxation of which muscle? _____

20. On the left-hand diagram, which direction does the diaphragm move? _____

21. On the left-hand diagram, which directions does the ribcage move? _____

22. What effect do the movements on the left-hand diagram have on lung volume? _____

23. On the right-hand diagram, which direction does the diaphragm move? _____

24. On the right-hand diagram, which directions does the ribcage move? _____

25. What effect do the movements on the right-hand diagram have on the lungs? _____

Dichotomous Choice

Circle one of two possibilities given between parentheses in each statement. [pp.184-185]

26. The air movements of respiration result from rhythmic increases and decreases in the volume of the (lungs / chest cavity).

27. When the air pressure in alveolar sacs is lower than atmospheric pressure, air flows down its gradient and (enters / exits) the alveoli.

28. The normally passive - does not require energy use - part of the respiratory cycle is (inspiration / expiration).

29. When the muscles that cause inhalation relax, air flows (into / out of) the lungs.

30. The negative pressure gradient outside the lungs keeps the lungs snug against the wall of the (pleural / chest) cavity even during exhalation.

31. The cohesiveness of water molecules in the fluid inside the (pleural sac / thoracic cavity) helps keep the lungs close to the thoracic wall.

32. Pneumothorax, or collapsed lung, is a condition in which air can enter the (intercostal / pleural) cavity, preventing normal expansion of the lungs.

33. The amount of air taken in by a person in a normal breath, about 500 ml, is known as the (vital capacity / tidal volume).

34. In addition to the air taken in as tidal volume, a person can forcibly inhale roughly 3,100 ml of air, called the (expiratory / inspiratory) reserve volume.

35. Forcibly exhaling, you can expel an additional (expiratory / inspiratory) reserve volume of about 1,200 ml.

36. The maximum volume of air that can move out of the lung after a person inhales as deeply as possible is called the (vital capacity / tidal volume).

37. People rarely take more than (one-half / three-fourths) of their vital capacity, even when they breathe deeply during strenuous exercise.

38. Even at the end of your deepest exhalation, your lungs still cannot be completely emptied of air; roughly another 1,200 ml of (expiratory reserve / residual) volume remains.

39. About (150 / 350) ml of inhaled air is actually available for gas exchange.

40. The Heimlich maneuver is used to dislodge food from the (trachea / esophagus).

10.5. HOW GASES ARE EXCHANGED AND TRANSPORTED [pp.186-187]

10.6. CONTROLS OVER BREATHING [pp.188–189]

10.7. RESPIRATORY SYSTEM DISORDERS: TOBACCO, IRRITANTS AND APNEA [pp.190–191]

10.8. PATHOGENS AND CANCER IN THE RESPIRATORY SYSTEM [p.192]

10.9. THE RESPIRATORY SYSTEM IN HOMEOSTASIS [p.193]

Boldfaced Terms

respiratory membrane _____

_____

infant respiratory distress syndrome _____

_____

oxyhemoglobin _____

_____

carbaminohemoglobin _____

_____

carotid bodies _____

_____

aortic bodies _____

_____

bronchitis_____

_____

emphysema _____

_____

asthma _____

_____

pneumonia _____

_____

influenza _____

_____

tuberculosis _____

_____

lung cancer _____

_____

squamous cell carcinomas _____

_____

adenocarcinomas _____

_____

large cell carcinomas _____

_____

small-cell carcinoma _____

_____

Matching

Choose the most appropriate description for each term. [pp.186-188]

1. ____ aortic bodies
2. ____ carbaminohemoglobin (HbCO$_2$)
3. ____ carbonic anhydrase
4. ____ carotid bodies
5. ____ external respiration
6. ____ infant respiratory distress syndrome
7. ____ internal respiration
8. ____ oxyhemoglobin (HbO$_2$)
9. ____ respiratory membrane
10. ____ carbon dioxide
11. ____ pulmonary surfactant
12. ____ medulla

A. Hemoglobin with bound oxygen

B. Enzyme in red blood cells mediating the chemical reactions that form and dissociate carbonic acid

C. Phase of respiration in which oxygen moves from blood into tissues, and carbon dioxide moves from tissues into blood

D. Sensory receptors located where carotid arteries branch to the brain; detect arterial changes in carbon dioxide and oxygen levels as well as pH

E. Formed by adjacent thin cell layers of alveoli and lung capillaries

F. Location of pacemaker for respiration

G. Life-threatening breathing disorder in premature infants whose partially developed lungs do not yet have functional surfactant-secreting cells.

H. Sensory receptors in arterial walls near the heart that detect changes in carbon dioxide, oxygen, and pH levels

I. Changes in levels of this compound cause the medulla to change respiratory rate

J. Carbon dioxide bound with hemoglobin

K. Phase of respiration in which oxygen moves from alveoli to blood and carbon dioxide moves from blood to alveoli

L. Secreted by certain cells of the alveolar epithelium; reduces surface tension of the watery fluid film between alveoli

Dichotomous Choice

Circle one of the two possibilities given between parentheses in each statement. [pp186-187]

13. Hemoglobin allows blood to carry (17 / 70) times more oxygen than it could otherwise.

14. The higher the partial pressure of oxygen around the alveoli, the (less / more) oxygen will be picked up by hemoglobin.

15. Oxygen-binding to hemoglobin weakens as temperature rises, or as pH (increases / decreases).

16. During (external/ internal) respiration, oxygen moves from blood into tissues.

17. About 70 percent of the body's carbon dioxide is transported in plasma in the form of (carbonic acid / bicarbonate).

18. Bicarbonate forms after carbon dioxide combines with (water / carbonic acid) in plasma.

19. Carbonic anhydrase mediates reactions that convert unbound carbon dioxide to carbonic acid and its dissociation products; the blood level of carbon dioxide then (rises / falls) rapidly.

20. In the alveoli, the partial pressure of carbon dioxide is (higher / lower) than it is in the surrounding capillaries, so that it diffuses into the sacs and is exhaled.

21. The H⁺ formed in red blood cells along with bicarbonate binds to hemoglobin which acts as a (buffer / plasma protein) to minimize pH change.

Fill-in-the-Blanks [p.188]

The body's control over breathing magnitude involves monitoring the level of (22) _____ in the blood. However, the nervous system is more sensitive to levels of (23) _____ . Both oxygen and carbon dioxide levels are monitored in blood flowing through (24) _____ . When conditions warrant, nervous system signals adjust contractions of the (25) _____ and muscles in the chest wall, and so adjust the rate and depth of breathing. Sensory receptors in the (26) _____ of the brain detect rising carbon dioxide levels. The receptors detect hydrogen ions in the (27) _____ fluid, which bathes the medulla. The shift in pH stimulates the receptors that signal changes to the brain's (28) _____ centers. The rate and depth of breathing fall, followed by a drop in the blood level of (29) _____ . In addition, the brain receives input from sensory receptors such as the (30) _____ bodies located where the carotid arteries branch to the brain, and the (31) _____ bodies in arterial walls near the heart. Both types of receptors can detect changes in carbon dioxide and oxygen levels in arterial blood as well as changes in blood pH. The brain responds by increasing the (32) _____ rate, so more oxygen can be delivered to tissues.

Dichotomous Choice

Circle one of the two possibilities between parentheses in each statement. [p.189]

33. If blood flow is too fast compared to ventilation rate, there is an a(n) (increase / decrease) in the blood level of carbon dioxide.

34. An increase in the blood level of carbon dioxide affects smooth muscle in the bronchiole walls so that the bronchioles (constrict / relax), increasing air flow.

35. A decrease in the blood level of carbon dioxide causes the bronchiole walls to (constrict / relax), so air flow decreases.

36. The mechanisms by which the nervous system regulates respiration normally operate under (voluntary / involuntary) control.

Fill-in-the-Blanks [p.190]

Cigarette smoke—including (37) "_____ smoke" inhaled by a nonsmoker—causes lung

cancer and contributes to various other ills. Cigarette smoking causes at least (38) _____] percent of all

lung cancer deaths.

Consider Your Own Health

39. A considerable amount of scientific research has been reported that firmly establishes the devastating effects of smoking on human beings. Study the table on page 206 closely, and then complete the following table by first checking in the left column only those risks associated with smoking that *you personally are willing to take* as a smoker. Be prepared to seriously discuss with your friends, relatives, and classmates the reasons you have for accepting the risks you checked. Then, in the right column, check which risks you are willing to subject nonsmokers among your family and close friends to through secondhand smoke. [p.190]

Personal Risks	Risks Associated with Smoking	Risks to Others
	a. shortened life expectancy	
	b. chronic bronchitis, emphysema	
	c. lung cancer	
	d. cancer of the mouth	
	e. cancer of the larynx	
	f. cancer of the esophagus	
	g. cancer of the pancreas	
	h. cancer of the bladder	
	i. coronary heart diseases	
	j. effects on your offspring	
	k. impaired immune system	
	l. slow bone healing (about 30% longer)	

Matching

Choose the most appropriate description for each term. [pp.190-191]

40. ____ asthma

41. ____ influenza

42. ____ sleep apnea

43. ____ emphysema

44. ____ bronchitis

45. ____ pneumonia

46. ____ tuberculosis

A. Lungs become so distended and inelastic that gas exchange is increasingly inefficient

B. Inflamation in lung tissue due to infection; fluid build-up makes breathing difficult

C. Serious lung infection that destroys patches of lung tissue; caused by bacteria that are becoming antibiotic-resistant

D. Inflammation of bronchial tree; may be chronic if constantly exposed to air pollution or cigarette smoke

E. Viral infection that beings in nose or throat and spreads to lungs; may trigger pneumonia

F. Bronchioles suddenly narrow due to strong spasms of smooth muscles in their walls

G. Breathing that stops briefly and then resumes, more common with aging

Short Answer

47. Explain the role of the respiratory system in homeostasis. [p.193] _____

_____

_____

_____

_____

SELF-QUIZ

____ 1. Most forms of life depend on _____ down concentration gradients to obtain oxygen and eliminate carbon dioxide. [p.182]

    a. active transport
    b. bulk flow
    c. diffusion
    d. osmosis
    e. muscular contractions

____ 2. _____ is the most abundant gas in Earth's atmosphere. [p.182]

    a. Water vapor
    b. Oxygen
    c. Carbon dioxide
    d. Hydrogen
    e. Nitrogen

___ 3. _____ are involved in local chemical controls over air flow operation in the lungs. [p.189]

a. Smooth muscles in bronchiole walls
b. Nerve cell clusters in the pons and medulla
c. The vagus nerve and its associated stretch receptors
d. Carotid bodies
e. Aortic bodies

___ 4. The amount of a gas diffusing across a respiratory surface does not depend on the _____. [pp.182,186]

a. amount of surface area of the membrane involved
b. level of glucose in the blood
c. differences in partial pressures of a gas across the membrane involved
d. presence and amount of hemoglobin
e. whether or not a respiratory membrane is moist

___ 5. Immediately before reaching the alveoli, air passes through the _____ [p.181].

a. bronchioles
b. glottis
c. larynx
d. pharynx
e. trachea

___ 6. During inhalation, _____ [p.184].

a. air pressure in the alveolar sacs is lower than atmospheric pressure
b. fresh air follows a gradient from the alveoli up to the trachea
c. the diaphragm moves upward and becomes more curved
d. the thoracic cavity volume decreases
e. all of the above

___ 7. Hemoglobin _____ [pp.186-187].

a. releases oxygen more readily in active tissues
b. tends to release oxygen in places where the temperature is lower
c. tends to hold on to oxygen when the pH of the blood drops
d. tends to give up oxygen in regions where partial pressure of oxygen is higher than in the blood
e. all of the above

___ 8. Because mechanisms for sensing carbon dioxide and oxygen levels become less effective, along with loss of lung elasticity with age, older people are more affected by _____ [p.191].

a. sleep apnea
b. emphysema
c. asthma
d. bronchitis
e. edema

___ 9. The wall of an alveolus separated from lung capillaries by a thin film of interstitial fluid constitutes a _____ [p.186].

a. bronchial tree
b. pleural sac
c. respiratory membrane
d. pulmonary surfactant
e. respiratory center

___ 10. Which of these is not known to be a risk of smoking cigarettes? [p.190]

a. negative effects on newborn babies
b. delayed healing of bones
c. impaired immunity
d. heart disease
e. tuberculosis

CHAPTER OBJECTIVES/REVIEW QUESTIONS

1. List all the principal parts of the human respiratory system and explain how each structure contributes to transporting oxygen from the external world to the bloodstream. [pp.180–181]

2. Describe the behavior of gases and the type of respiratory surface that participates in gas exchange in humans. [p.182]

3. The two phases of ventilation are _____ and _____. [p.184]

4. Describe the functional relationship of the human lung to the chest cavity. [p.184]

5. For the human lung, distinguish tidal volume from vital capacity. [p.185]

6. Distinguish between hemoglobin, oxyhemoglobin and carbaminohemoglobin. [pp.186-187]

7. Explain why oxygen diffuses from the bloodstream into the tissues far from the lungs. Then explain why carbon dioxide diffuses into the bloodstream from the same tissues. [pp.186–187]

8. Explain why oxygen diffuses from alveolar air spaces, through interstitial fluid, and across capillary epithelium. Then explain why carbon dioxide diffuses in the reverse direction. [pp.186–187]

9. Describe what happens to carbon dioxide when it dissolves in water under conditions normally present in the human body. [p.187]

10. List the structures involved in detecting carbon dioxide levels in the blood and in regulating the rate of breathing. Name the location of each structure. [pp.188–189]

11. List seven respiratory disorders or diseases, and the characteristics of each. [pp.190-192]

INTEGRATING AND APPLYING KEY CONCEPTS

1. Explain the need for an extensive and efficient circulatory system to go along with the human respiratory system, including the huge number of capillaries, efficient heart, and presence of hemoglobin.

2. Consider some of the respiratory diseases related to airborne substances in the environment. What methods might be implemented to reduce the risk of contracting these diseases?

11

DIGESTION AND NUTRITION

INTRODUCTION

After examining the organs of the digestive system and their roles, diet and nutrition are discussed. Considering the fact that health problems associated with obesity are becoming increasingly common in the United States, an understanding of digestion and nutrition is certainly practical.

FOCAL POINTS

- Figure 11.1 animated [p.198] diagrams the parts of the gastrointestinal tract as well as accessory organs of digestion.
- Figure 11.6 animated [p.201] demonstrates the components of the process of swallowing.
- Figure 11.8 [p.203] presents the folds within folds structure of the small intestine. The concept of increased surface area by folding recurs frequently in the study of Biology.
- Table 11.1 animated [p.207] illustrates absorption of nutrients in the small intestine.
- Tables 11.4 [p.216] and 11.5 [p.217] provide a good summary of the sources, functions and deficiencies of various vitamins and minerals.

INTERACTIVE EXERCISES

CHAPTER INTRODUCTION [p.197]

11.1. OVERVIEW OF THE DIGESTIVE SYSTEM [pp.198–199]

11.2. CHEWING AND SWALLOWING: FOOD PROCESSING BEGINS [pp.200–201]

11.3. THE STOMACH: FOOD STORAGE, DIGESTION, AND MORE [p.202]

Boldfaced Terms

digestive system _____

_____

sphincters_____

_____

mechanical processing_____

_____

motility _____

_____

secretion _____

_____

digestion _____

_____

absorption _____

_____

elimination _____

_____

salivary glands _____

_____

salivary amylase _____

_____

bolus _____

_____

palate _____

_____

pharynx _____

_____

esophagus _____

_____

peristalsis _____

_____

pepsins _____

_____

gastric juice _____

_____

166   Chapter Eleven

chyme _____

_____

rugae _____

_____

Labeling

Identify the numbered parts of the accompanying illustration. [p.198]

1. _____

2. _____

3. _____

4. _____

5. _____

6. _____

7. _____

8. _____

9. _____

10. _____

11. _____

12. _____

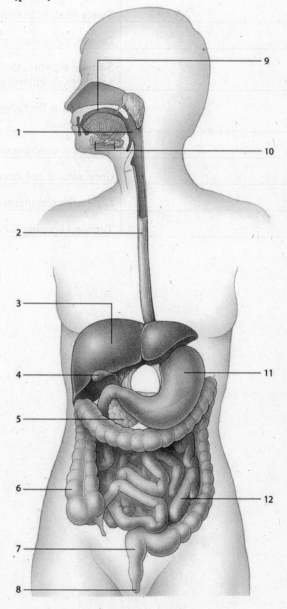

Complete the Table

13. Complete the following table by naming the digestive system components described. [p.198]

Organ	Main Functions
a.	Food is moistened and chewed; polysaccharide digestion starts
b.	Secrete saliva containing digestive enzymes, buffers and mucus
c.	Passageway to both the tubular part of the digestive system and to the respiratory system; moves food forward by contracting sequentially
d.	Muscular tube, moistened by saliva, which moves food from the pharynx to the stomach
e.	Stores food; kills many microorganisms; starts protein digestion
f.	Digests and absorbs most nutrients
g.	Secretes enzymes that break down all major food molecules; produces buffers against hydrochloric acid from stomach
h.	Secretes bile for fat emulsification; roles in metabolism of carbohydrates, fats, and proteins
i.	Stores and concentrates bile produced by liver
j.	Concentrates and stores undigested matter by absorbing ions and water
k.	Distension stimulates expulsion of feces
l.	Terminal opening for expelling feces

Labeling and Matching

First, identify the numbered parts on the diagram of the section of the gastrointestinal tract shown below. Then, match each component to its lettered description to the right. [p.199]

14. _____

15. _____

16. _____

17. _____

18. _____

A. connective tissue with blood and lymph vessels and nerve cells

B. thin serous membrane

C. space through which food passes

D. consists of two sublayers, circular and longitudinal

E. innermost layer of epithelium, coated with mucus

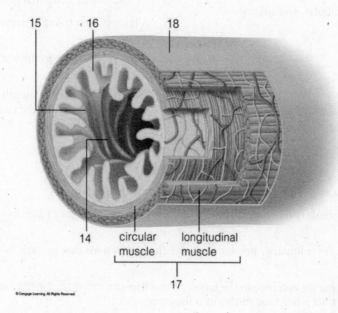

15   16   18

14   circular   longitudinal
     muscle   muscle

17

© Cengage Learning. All Rights Reserved.

Short Answer

19. Describe the structure and function of a sphincter muscle. [p.199] _____

_____

_____

_____

20. In addition to the components of the gastrointestinal (GI) tract, the digestive system includes accessory structures. List four of these. [p.198] _____

_____

_____

_____

Digestion and Nutrition   **169**

Matching

Select the description that best matches the numbered term. [pp.200-201]

21. _____ bolus
22. _____ enamel
23. _____ esophagus
24. _____ premolars/molars
25. _____ oral cavity
26. _____ palate
27. _____ pulp cavity
28. _____ pharynx
29. _____ salivary amylase
30. _____ parotid, submandibular, and sublingual

A. Part of tooth containing blood vessels and nerves

B. A muscular tube leading from pharynx to stomach

C. A softened ball of food formed by activity of mucins

D. A term synonymous with throat

E. Types of teeth used to grind up food

F. Three pairs of saliva-producing glands in the vicinity of the ears, lower jaw, and under the tongue

G. An enzyme in saliva that breaks down starch

H. A term synonymous with mouth; where chewing occurs

I. Hardest substance in the body; coats the crown of a tooth

J. Hardened roof of the mouth providing a hard surface against which the tongue can press food as it mixes with saliva

Dichotomous Choice

Circle one of two possible answers given between parentheses in each statement. [p.201]

31. Swallowing begins when voluntary movements push a bolus into the (esophagus / pharynx) stimulating sensory receptors in its wall.

32. The sensory receptors trigger (voluntary / involuntary) muscle contractions that prevent food from entering the nose and the trachea.

33. The vocal cords stretch across the entrance to the larynx, and a flap-like valve, the (uvula / epiglottis), closes the opening to the respiratory tract while food moves into the esophagus.

34. To enter the stomach, food must pass through a (sphincter / gland).

Fill-in-the-Blanks [p.202]

The (35) _____ is a muscular, stretchable sac that stores food, helps break it down, and controls its passage into the (36) _____ . Gland cells in the stomach epithelium produce (37) _____ (HCl), mucus, and pepsinogens which are precursors of digestive enzymes known as (38) _____ . Gland cells in the stomach lining also secrete (39) _____ factor , a protein necessary for vitamin B$_{12}$ absorption in the small intestine. All of these substances combined are referred to as (40) _____ . Combined with stomach contractions, the acidity in the stomach helps convert swallowed boluses into a thick mixture called (41) _____ .

Digestion of (42) _____ begins in the stomach. The high acidity due to HCl secretion (43) _____ proteins and exposes their peptide bonds. The acidity also converts inactive pepsinogens to active (44) _____ , which break the bonds of the long protein molecules. The secretion of HCl and pepsinogen is stimulated by the hormone (45) _____ that is secreted from gland cells.

The (46) "_____" of mucus and bicarbonate prevents the HCl and pepsin from breaking down the inner surface of the stomach lining.   In the stomach, waves of contraction and relaxation called (47) _____ mix the chyme and build up force as they approach the (48) _____ sphincter. Strong contractions of the stomach wall (49) _____ the sphincter. This squeezes most of the chyme back so that only a small amount moves into the (50) _____ with each contraction. From two to six hours later, the stomach is empty and its walls crumple into folds called (51) _____. Only (52) _____, (53) _____ and a few other substances begin to be absorbed across the stomach wall.

11.4. THE SMALL INTESTINE: A HUGE SURFACE FOR DIGESTION AND ABSORPTION [p.203]

11.5. ACCESSORY ORGANS: THE PANCREAS, GALLBLADDER, AND LIVER [p.204-205]

11.6. DIGESTION AND ABSORPTION IN THE SMALL INTESTINE [p.206-207]

11.7. THE LARGE INTESTINE [p.208]

Boldfaced Terms
villus (plural: villi) _____

_____

microvillus_____

_____

brush border _____

_____

pancreas_____

_____

bile _____

_____

gallbladder _____

_____

liver _____

_____

hepatic portal vein _____

_____

duodenum _____

_____

jejunum _____

_____

ileum _____

_____

segmentation _____

_____

lacteals _____

_____

colon _____

_____

rectum _____

_____

appendix _____

_____

anal canal _____

_____

anus _____

_____

Fill-in-the-Blanks [p.203]

The key to the ability of the small intestine's ability to absorb nutrients is the structure of its (1)

_____ . The mucosa of the small intestine is densely folded, tremendously increasing the (2)

_____ available for absorbing nutrients. The folds have smaller hairlike projections called (3)

_____ . Small (4) _____ in each villus move substances to and from the (5)

_____ . The epithelial cells covering each villus usually have threadlike projections of the plasma

membrane called (6) _____ . This dense array gives the epithelium of villi its common name, the "(7)

_____ " . (8) _____ in the lining release digestive enzymes.

Choice

For questions 9-17, determine the accessory organ below to which the statement applies. [pp.204-205]

    a. pancreas   b. liver   c. gall bladder

9. ____ along with digestive functions, this large organ also removes toxins from the blood

10. ____ a small sac found behind the liver

11. ____ secretes a buffering fluid containing bicarbonate ions

12. ____ releases bile into small intestine

13. ____ slender organ located behind and below the stomach

14. ____ produces bile

15. ____ makes and secretes digestive enzymes

16. ____ uses cholesterol to make bile salts

17. ____ precipitation of excess cholesterol in this dispensable organ causes gallstones

Dichotomous Choice

Circle one of two possible answers given between parentheses in each statement. [p.205]

18. Nutrient-laden blood from the small intestine flows to the (hepatic portal vein / hepatic vein).

19. In the liver, excess glucose is taken up before a (hepatic portal vein / hepatic vein) returns the blood to the general circulation.

20. The liver converts and stores much of the glucose to (glycogen / glucagon).

21. The liver removes (fats / toxins) from the bloodstream.

22. Toxic ammonia (NH_3) produced when cells break down amino acids is carried by the circulatory system to the liver, where it is converted to less toxic (glucose / urea) that is excreted in the urine.

Choice

Choose the location(s) from the choices below where digestion of the following molecules occurs.. [p.207]

    a. mouth   b. stomach   c. small intestine

23. ____, *____, ____ - carbohydrates (polysaccharides, disaccharides)

                       *digestion in stomach continues only until salivary amylase is denatured by acid

24. ____, ____ - proteins (and peptides)

25. ____ - fats (triglycerides)

26. ____ - nucleic acids

Choice

Choose the organ that secretes the enzyme(s) listed from the choices below. [p.207]

    a. mouth   b. stomach   c. small intestine   d. pancreas

27. ____ - trypsin, chymotrypsin

28. ____ - salivary amylase

29. ____ - lipase

30. ____ , ____ - nucleases

31. ____ - pancreatic amylase

32. ____ - pepsin

33. ____ , ____ - peptidases

Short Answer

34. What characteristic of fats makes them difficult to digest? What role do bile salts play in emulsifying fat? [p.206]

_____

_____

_____

_____

35. During digestion in the small intestine, what is the value of the process called segmentation? [p.206]

_____

_____

_____

_____

Fill-in-the-Blanks [p.206]

By the time food is halfway through the (36) _____ , most of it has been digested. Water

crosses the intestine lining by (37) _____ , and certain minerals are absorbed. (38) _____

in the plasma membrane of brush border cells actively move nutrients like (39) _____ and (40)

_____ across the intestine lining and directly into the bloodstream. Fat globules are more difficult to

absorb. Molecules of fat clump with (41) _____ , forming tiny droplets called (42) _____ .

Nutrients diffuse out of micelles into epithelial cells where triglycerides are reformed. Instead of moving directly into

the bloodstream, triglycerides enter lymph vessels called (43) _____ .

Matching

Choose the most appropriate description for each term.

44. _____ anal canal

45. _____ appendix

46. _____ cecum

47. _____ *Escherichia coli*

48. _____ feces

49. _____ colon

50. _____ rectum

51. _____ anus

A. Terminal end of gastrointestinal tract; sphincter muscle controls elimination of feces

B. Connects with the sigmoid colon; feces here move toward the outside of the body through the anal canal

C. A mixture of undigested and unabsorbed food material, water, and bacteria

D. Receives material not absorbed in the small intestine

E. A blind pouch that is the beginning of the large intestine

F. A slender projection from the cecum with no known digestive function

G. Feces within this structure are moving directly toward the outside of the body

H. Normally inhabit intestines for nourishment but also furnish useful fatty acids and vitamins, such as vitamin K

11.8. CONTROLS OVER DIGESTION [p.209]

11.9. DIGESTIVE SYSTEM DISORDERS [p. 210-211]

11.10. INFECTIONS IN THE DIGESTIVE SYSTEM [p. 212]

11.11. THE DIGESTIVE SYSTEM IN HOMEOSTASIS [p. 213]

Matching

Choose the most appropriate description for each term. [p.209]

1. _____ cholecystokinin (CCK)

2. _____ gastrin

3. _____ glucose insulinotropic peptide

4. _____ secretin

5. _____ somatostatin

A. Hormone that stimulates the pancreas to secrete bicarbonate

B. Gastrointestinal hormone released in response to the presence of glucose and fat in the small intestine; stimulates insulin release

C. Gastrointestinal hormone that inhibits acid secretion

D. Gastrointestinal hormone released in response to fat in the small intestine; enhances the actions of secretin and stimulates gallbladder contractions

E. Hormone that stimulates the secretion of acid into the stomach in the presence of peptides and amino acids

Matching

Choose the most appropriate description for each term. [pp.210-212]

6. _____ bulk
7. _____ constipation
8. _____ diarrhea
9. _____ insoluble fiber
10. _____ Crohn's disease
11. _____ irritable bowel syndrome
12. _____ colorectal cancer
13. _____ soluble fiber
14. _____ cystic fibrosis
15. _____ lactose intolerance
16. _____ food poisoning
17. _____ gingivitis
18. _____ caries
19. _____ peptic ulcers
20. _____ gastritis
21. _____ polyp
22. _____ diverticulitis
23. _____ diverticulosis
24. _____ gastroesophageal reflux disease

A. feces become dry and hard, and defecation becomes difficult

B. the volume of fiber and other undigested food material that cannot be decreased by absorption in colon

C. tooth decay

D. inflammation of a knoblike sac where inner colon lining protrudes through intestine wall

E. plant carbohydrates that swell or dissolve in water

F. occurs when an irritated mucosal lining secretes more water and salts than the large intestine can absorb

G. cellulose and other plant compounds that cannot be digested by humans

H. growth on colon wall that may become malignant

I. open sores in the wall of the stomach or small intestines

J. due to disturbance in smooth muscle contractions in colon; cause of disturbance is unknown

K. sufferers do not produce pancreatic enzymes necessary for normal digestion and absorption of fats and other nutrients

L. a malabsorption disorder resulting from deficiency of the enzyme lactase

M. inner colon has protruded through intestine wall but no inflammation has developed

N. inflammation of the GI tract

O. results from colonization of stomach or intestines with bacteria such as Salmonella or E. coli from contaminated food

P. develops from polyp; #2 cancer diagnosis in U.S.

Q. disorder that causes such severe damage to the intestinal lining that much of intestine is removed

R. acidic chyme backs up into the esophagus

S. inflammation of the gums that affects periodontal membrane

Matching

Choose the most appropriate description for each term. [pp.210-212]

25. ____ hepatitis
26. ____ malabsorption disorder
27. ____ alcoholic cirrhosis
28. ____ rotavirus
29. ____ gluten intolerance
30. ____ giardiasis

A. scarred liver due to heavy alcohol consumption
B. celiac disease; hypersensitivity to wheat
C. protozoan in water or food that causes diarrhea
D. inflammation of the liver
E. inability to absorb nutrients
F. common viral cause of diarrhea

Short Answer

31. Explain how the digestive system contributes to homeostasis. [p.213]_____

_____

_____

_____

_____

11.12. THE BODY'S NUTRITIONAL REQUIREMENTS [pp.214–215]

11.13. VITAMINS AND MINERALS [pp.216-217]

11.14. FOOD ENERGY AND BODY WEIGHT [pp.218–219]

11.15. DEALING WITH WEIGHT EXTREMES [pp.220-221]

Boldfaced Terms

glycemic index _____

_____

essential fatty acids _____

_____

essential amino acids _____

_____

vitamins _____

_____

minerals _____

_____

phytochemicals _____

_____

BMI _____

_____

obesity _____

_____

kilocalories _____

_____

basal metabolic rate (BMR) _____

_____

appetite _____

_____

metabolic syndrome _____

_____

binge eating _____

_____

anorexia nervosa _____

_____

bulimia nervosa _____

_____

Choice

Determine the class of nutrient to which each statement refers. Use these choices. [pp.214-215]

  a. carbohydrates   b. fats   c. proteins

1. _____   The liver can make most of these required by the body.

2. _____   "Complex" ones include fruits, grains and legumes.

3. _____   Amino acids from these foods become available for synthesis within our cells.

4. _____   The main use of these is as the body's chief energy source.

5. _____   Incomplete forms must be combined to provide all essential amino acids.

6. _____   Those from animals are associated with heart disease, stroke and some cancers.

7. _____   Products of digestion of this food type are taken in by cells with the help of insulin.

8. _____   This group includes phospholipids and sterols such as cholesterol.

9. _____   Complete forms include meat and fish.

10. _____   Excess is stored in adipose tissue.

Short Answer

11. Why are simple sugars classified as "empty calories"? [p.214] _____

    _____

    _____

    _____

12. What does the term "glycemic index" refer to? [p.214] _____

    _____

    _____

    _____

13. What do essential fatty acids and essential amino acids have in common? [p.214] _____

    _____

    _____

    _____

14. In what two ways should Americans change their diet, according to the nutritional guidelines issued by the U.S. Food and Drug Administration? [p.215] _____

    _____

    _____

    _____

Dichotomous Choice

Circle one of two possible answers given between parentheses in each statement. [pp.216-217]

15. Vitamins are (organic / inorganic) while minerals are (organic / inorganic).

16. We must get vitamins from foods because our cells don't synthesize (enough / any) of these essential molecules.

17. People in good health probably (do / do not) get enough vitamins and minerals from a balanced diet of whole foods.

18. Vitamins A, C and E retard aging and improve immune function by (promoting / inhibiting) free radicals in our bodies.

19. Excessive intake of vitamins (is / is not) harmful.

20. Vitamins A, D, E and K are (fat- / water-) soluble.

21. Thiamine and riboflavin are types of (B / C) vitamins .

22. Ascorbic acid is another name for Vitamin (B_{12} / C).

23. The B vitamins are essential as a source of (hormones / coenzymes) that help enzymes to function.

Matching

Choose the most appropriate description for each term. [pp.218-219]

24. _____ BMI that defines obesity

25. _____ sedentary people multiply weight by this number to determine the kilocalories that they need

26. _____ 1kilocalorie = _____ calories

27. _____ losing a pound requires expending this number of calories

A. 1000

B. 10

C. 30

D. 3500

Complete the Table

28. Complete the following table by determining how many kilocalories the people described should take in daily, given the stated exercise level, to maintain their weight. [pp.218-219]

Height	Age	Sex	Level of Physical Activity	Present Weight	Number of Kilcalories
5'1"	25	Female	Moderately active	140	
6'2"	37	Male	Very active	140	
5'8"	59	Female	Not very active	140	

Fill in the Blanks

Use the BMI chart, Table 11.6, [p.218] of the textbook to answer the following questions.

29. Which of the individuals from the chart above is overweight? _____

30. Which is underweight? _____

31. Which is at the ideal weight? _____

32. How many hours would the 25-year-old female have to jog to lose one pound? _____

Short Answer

33. List some health risks associated with obesity. [p.220] _____

_____

_____

_____

34. Distinguish between the eating disorders anorexia nervosa and bulimia. [p.221] _____

_____

_____

_____

_____

SELF-QUIZ

Multiple Choice

_____ 1. What hormone secreted by adipose cells causes a person to feel "full"? [p.219]

    a. ghrelin
    b. endorphin
    c. calcitonin
    d. thyroxine
    e. leptin

_____ 2. What percentage of North Americans are overweight or obese? [p.197]

    a. 10%
    b. 25%
    c. 45%
    d. 66%
    e. 90%

_____ 3. The process that moves nutrients into the blood or lymph is _____ . [p.199]

    a. ingestion
    b. absorption
    c. assimilation
    d. digestion
    e. none of the above

_____ 4. The enzymatic digestion of proteins begins in the _____. [p.202]

    a. mouth
    b. stomach
    c. liver
    d. pancreas
    e. small intestine

_____ 5. The enzymatic digestion of starches begins in the _____. [p.200]

    a. mouth
    b. stomach
    c. liver
    d. pancreas
    e. small intestine

_____ 6. A bolus moves from the pharynx to the _____. [p.201]

    a. trachea
    b. larnyx
    c. glottis
    d. oral cavity
    e. esophagus

7. What is the main force in moving food through the gastrointestinal tract? [p.201]

a. cilia
b. peristalsis
c. sphincters
d. swallowing
e. osmosis

8. Digestion of fats requires bile and _____. [p.206]

a. lecithin
b. cholesterol
c. pancreatic enzymes
d. pigments
e. *Escherichia coli*

9. Water moves through the membranes of the small intestine by _____. [p.206]

a. peristalsis
b. osmosis
c. diffusion
d. active transport
e. bulk flow

10. Which one of the following does not apply to the large intestine? [p.208]

a. It contains large populations of bacteria.
b. It is divided into the duodenum, jejunum, and ileum.
c. It concentrates undigested matter.
d. It stores undigested matter.
e. It absorbs water.

11. Males tend to burn more kilocalories than females because men _____. [p.219]

a. are more active
b. are taller
c. weigh more
d. have more muscle
e. have a higher BMR

12. _____ is an eating disorder in which an individual purposely starves and overexercises. [p.221]

a. Malabsorption disorder
b. Peritonitis
c. Bulimia
d. Anorexia nervosa
e. Diverticulosis

13. Which of these would most Americans have to decrease if they tried the Mediterranean Diet? [p.215]

a. carbohydrates
b. red meat
c. oil
d. nucleic acids
e. amino acids

Matching

14. ____ gallbladder [p.198]
15. ____ large intestine [p.198]
16. ____ liver [p.198]
17. ____ oral cavity [p.198]
18. ____ pancreas [p.198]
19. ____ small intestine [p.198]
20. ____ stomach [p.198]

A. Secretes bile
B. Secretes an enzyme for each major food category; secretes insulin
C. Where most digestion and absorption occurs
D. Where water and salts are absorbed, where indigestible food is concentrated and stored
E. Stores bile
F. Holds food temporarily, secretes ghrelin
G. Where salivary amylase works

CHAPTER OBJECTIVES/REVIEW QUESTIONS

1. List all specialized regions (in order) of the human gastrointestinal tract through which food actually passes. Then list the accessory structures that contribute one or more substances to the digestive process. [pp.198–208]

2. Define and distinguish among *mechanical processing and motility, secretion, digestion, absorption,* and *elimination.* [p.199]

3. Briefly describe the four-layered wall of the gastrointestinal tract. [p.199]

4. Describe mechanical and chemical digestive processes occurring in the mouth. [p.200]

5. What role do bacteria play in causing stomach ulcers? [p.212]

6. Describe the role of the stomach's acidity in protein digestion. [p.202]

7. Describe the cross-sectional structure of the small intestine, and explain how its structure is related to its function. [p.203]

8. What functions do the pancreas, gallbladder and liver play in digestion? [pp.204-205]

9. Describe how the digestion and absorption of fats differ from the digestion and absorption of carbohydrates and proteins. [pp. 206-207]

10. List the enzyme(s) that act in (a) the mouth, (b) the stomach, and (c) the small intestine. Then tell where each enzyme was originally produced. [p.207]

11. Tell which foods undergo digestion in each of the following parts of the human digestive system, and state what kinds of simple biological molecules they are broken into: oral cavity, stomach, small intestine. [p.207]

12. List the molecules that leave the digestive system and enter the circulatory system during the process of absorption. [p.207]

13. Summarize the processes that occur in the colon. [p.208]

14. How is the endocrine system involved in digestion? [p.209]

13. What causes constipation? diarrhea? [p.210, 212]

14. Compare the roles of carbohydrates, proteins, and fats in human nutrition to the roles of vitamins and minerals. [pp.214-217]

15. Distinguish vitamins from minerals. [p.216]

16. Name two minerals that are important in human nutrition and state the specific role of each. [p.217]

17. What type of body shape is most associated with health problems? What is the usual culprit in acquiring this overweight body shape? [p.218]

INTEGRATING AND APPLYING KEY CONCEPTS

1. Based on the current world distribution of basic foods versus human populations, suggest ways to prepare for an almost doubling of the world population within the next three decades. Include mention of how nutrition, not just food quantity, is relevant.

2. Why is it important for vegetarians to eat combinations of grains and legumes in order to stay healthy?

12

THE URINARY SYSTEM

INTRODUCTION

This chapter deals with the multiple functions of the urinary system. After studying the content, you should be able to describe the means by which solute and water excretion is regulated. It is very important, throughout the chapter, to consider the role of the urinary system in homeostasis.

FOCAL POINTS

- Figure 12.2 [p.227] shows the interaction between the urinary system and other organ systems.
- Figure 12.3 [p.228] is a diagram of the parts of the urinary system. Note the close association of the urinary system and the circulatory system. Being aware of the interactions of various body systems is critical to a useful understanding of the human organism.
- Figure 12.5 animated [p.230] provides a diagrammatic overview of urine formation. Again, notice the interaction of the urinary and circulatory systems.
- Figures 12.8 [p.232] and 12.10 [p.233] show the interaction of the nephron and regulatory hormones.
- Figure 12.12 animated [p.235] illustrates the processes of hemodialysis and peritoneal dialysis.

INTERACTIVE EXERCISES

CHAPTER INTRODUCTION [p.225]

12.1. THE CHALLENGE: SHIFTS IN EXTRACELLULAR FLUID [pp.226–227]

12.2. THE URINARY SYSTEM: BUILT FOR FILTERING AND WASTE DISPOSAL [pp.228–229]

12.3 HOW URINE FORMS: FILTRATION, REABSORPTION, AND SECRETION [pp.230-231]

Boldfaced Terms

urinary excretion _____

_____

urine _____

_____

urea _____

_____

electrolytes _____

kidney _____

ureter _____

urinary bladder _____

urethra _____

nephrons _____

glomerulus _____

Bowman's capsule _____

proximal tubule _____

loop of Henle _____

distal tubule _____

peritubular capillaries _____

filtration _____

tubular reabsorption _____

tubular secretion _____

Short Answer

1. What is the basic task of the urinary system? Use the terms *extracellular fluid* and *intracellular fluid* in your answer. [p.226] _____

_____

_____

_____

Complete the Table

2. Complete the table by categorizing the movements of water and solutes. [pp.226-227]

Water Gain	Water Loss	Solute Gain	Solute Loss
a.	c.	g.	k.
b.	d.	h.	l.
----------	e.	i.	m.
----------	f.	j.	----------

True/False

If the statement is true, write a T in the blank. If the statement is false, make it correct by writing the word(s) in the blank that should take the place of the underlined word(s). [pp.226-227]

3. _____ When there is a water deficit in body tissues, the <u>kidneys</u> compel us to seek out water.

4. _____ Of the ways that water is lost, the body exerts the most control over <u>sweating</u>.

5. _____ Water loss that a person is not aware of, such as through evaporation from lungs or skin, is called "<u>insensible</u>".

6. _____ The <u>excretory</u> system brings oxygen into the blood, and respiring cells add carbon dioxide to it.

7. _____ The most abundant metabolic waste is <u>carbon dioxide</u>.

8. _____ All metabolic wastes besides carbon dioxide leave in the <u>feces</u>.

9. _____ <u>Uric acid</u> forms when cells break down nucleic acids.

10. _____ Ammonia is a product of <u>carbohydrate</u> metabolism.

11. _____ The liver combines ammonia and carbon dioxide to make <u>urine</u>.

12. _____ The kidneys regulate the balance of ions such as sodium, potassium, and calcium that are called <u>electrolytes</u>.

Labeling

Identify each indicated part of the accompanying illustrations. [p.228]

13. _____

14. _____

15. _____

16. _____

17. _____

18. _____

19. _____

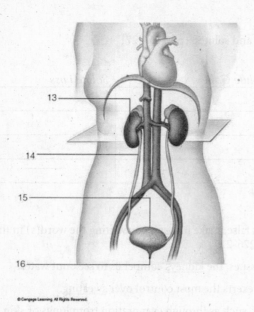

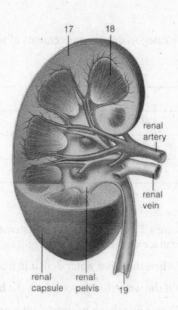

17
18

renal
artery

renal
vein

13

14

15

16

© Cengage Learning. All Rights Reserved.

renal
capsule

renal
pelvis

19

Short Answer

20. Name four functions of the kidneys. [p.228] _____

_____

_____

_____

_____

Fill-in-the-Blanks [pp.227-228]

　　　Each kidney lobe contains blood vessels and over a million slender tubes called (21) _____ that

filter water and (22) _____ from the (23) _____ . The wall of a nephron is a single layer of a

(24) _____ cells. Water and solutes pass easily through some parts of the wall, but other parts block

188　Chapter Twelve

solutes unless they are removed by (25) _____ .

The nephron balloons around a cluster of capillaries called the (26) _____ . Its cuplike wall region, the (27) _____ capsule, receives water and solutes filtered from the blood. This filtrate moves into the (28) _____ tubule, then through the loop of (29) _____ and from there into the (30) _____ tubule that empties into a collecting duct.

Blood is delivered to the kidneys by the (31) _____ . The blood flows into glomerular (32) _____ inside (Bowman's) capsule. Unlike other capillaries, these have slitlike pores that are especially (33) _____ , and they do not channel blood to venules, but converge to form a(n) (34) _____ arteriole. This branches into a set of (35) _____ capillaries that weave around the nephron's tubules. These then merge into (36) _____ that carry filtered blood out of the kidneys.

Labeling

Identify each indicated part of the accompanying illustration. [pp.229 - 232]

37. _____   41. _____

38. _____   42. _____

39. _____   43. _____

40. _____   44. _____

12.4. HOW KIDNEYS HELP MANAGE FLUID BALANCE AND BLOOD PRESSURE [pp.232-233]

12.5. REMOVING EXCESS ACIDS AND OTHER SUBSTANCES IN URINE [p.234]

12.6. KIDNEY DISORDERS: WHEN KIDNEYS FAIL [p.235]

12.7. CANCER, INFECTIONS, AND DRUGS IN THE URINARY SYSTEM [p.236]

12.8. THE URINARY SYSTEM IN HOMEOSTASIS [p.237]

Boldfaced Terms

ADH _____

_____

juxtaglomerular apparatus_____

_____

aldosterone_____

_____

acid-base balance _____

_____

metabolic acidosis _____

_____

metabolic alkalosis _____

_____

kidney stones _____

_____

glomerulonephritis _____

_____

polycystic kidney disease _____

_____

cystitis _____

_____

pyelonephritis _____

_____

nephritis_____

_____

Complete the Table

1. Complete the table below by categorizing events that occur in the kidneys into the three steps of urine formation. [pp.230-231]

Step 1: Filtration	Step 2: Reabsorption	Step 3: Secretion
a. occurs across _____ capillaries	d. most occurs across walls of _____ tubules	g. occurs along the _____ tubule
b. moves _____ and small _____ out of blood	e. most of _____ leaves or is pumped out of nephron tubule	h. unwanted substances are transported out of _____
c. forces substances into _____ capsule, then into _____ tubule of nephron	f. returns most of filtrate into _____ capillaries	i. substances are added to _____ forming in nephron tubules

Dichotomous Choice

Circle one of the two possible answers given between parentheses in each statement. [pp.230-231]

2. The force required for filtration is based on the small diameter of efferent arterioles that deliver blood to the glomerulus under (low / high) pressure.

3. During filtration, blood cells, platelets, proteins and other (small / large) substances remain in the blood.

4. During reabsorption, active transport moves glucose and Na^+ into the (bloodstream / tissue fluid).

5. Water follows solutes out of the proximal tubule into the peritubular capillaries by the process of (active transport / osmosis).

6. (Reabsorption / Secretion) is critical to maintaining the body's acid-base balance.

7. Secretion ensures that some wastes and foreign substances such as (uric acid / antibiotics) do not build up in the blood.

8. Drug testing relies on the use of (urinalysis / dialysis) which shows which substances have been secreted in the urine.

9. As well as removing impurities from the blood, the kidneys also adjust the amount of (urine / water) that is excreted or returned to the bloodstream.

10. Urination is a reflex response signaled by tension across the walls of the (kidney / urinary bladder).

11. During urination, the internal urethral sphincter (relaxes / contracts) while the bladder walls force urine through the urethra.

12. A person can exert control over the (internal / external) urethral sphincter.

True / False

If the statement is true, write a T in the blank. If the statement is false, make it correct by writing the word(s) in the blank that should take the place of the underlined word(s). [p.232]

13. _____ About half of filtered salt and water is reabsorbed in the proximal tubule.

14. _____ The loop of Henle descends into the kidney cortex

15. _____ Because the tissue fluid around the descending loop is very salty, water moves into the loop by osmosis.

16. _____ Water cannot pass through the ascending loop of Henle.

17. _____ Sodium is actively transported out of the descending loop of Henle

18. _____ In the distal tubule, water is removed from the filtrate.

19. _____ Some urea diffuses out of the final portion of the collecting duct, adding to the concentration of solutes in the outer medulla.

20. _____ The hormonal controls that allow the kidneys to regulate the amount of water in the urine also adjust blood pressure.

Dichotomous Choice

Circle one of the two possible answers given between parentheses in each statement. [pp232-233]

21. The release of antidiuretic hormone (ADH) is controlled by the (kidney / brain).

22. ADH acts on (proximal / distal) tubules and collecting ducts in the kidney cortex.

23. ADH causes (less / more) water to be reabsorbed from urine.]

24. A diuretic promotes the (loss / retention) of water in the urine.

25. A thirst center in the brain can inhibit the production of (ADH / saliva), making you feel thirsty.]

Sequencing

26. Place the numbers 1 through 4 in the blanks to indicate the order of these events controlling urine concentration. [pp.233]

____ Aldosterone acts on distal tubules preventing loss of sodium and water in urine.

____ Angiotensin I is converted to angiotensin II.

____ Aldosterone is secreted from adrenal cortex.

____ Renin is released from cells in the juxtaglomerular apparatus.

Fill-in-the-Blanks [p.234]

Normal pH of extracellular body fluids ranges from (27) _____ to _____ . Buffer

systems have a temporary effect in neutralizing excess H^+, but only the urinary system can (28) _____

excess H^+. Depending on changes in the blood's acid-base balance, the kidneys can either excrete (29)

_____ or form new bicarbonate and add it to (30) _____ . The chemical reactions involved

occur in the cells of (31) _____ walls.

Bicarbonate produced in these reactions moves into (32) _____ and ends up circulating in

the blood where it buffers excess (33) _____ . When the blood is too (34) _____ , chemical

adjustments in the kidneys ensure that (35) _____ bicarbonate is reabsorbed into the bloodstream.

H^+ formed in the tubule cells is secreted into the filtrate where it is combined with phosphate ions, (36)

_____ or bicarbonate. In this way, excess H^+ is (37) _____ .

Matching

Choose the most appropriate description for each term. [pp.235-236]

38. ____ polycystic kidney disease
39. ____ kidney stones
40. ____ nephritis
41. ____ glomerulonephritis
42. ____ cystitis
43. ____ hemodialysis
44. ____ lithotripsy

A. Bladder infection

B. Inflammation of the kidneys, interferes with blood filtering

C. Damage to kidneys from hypertension, diabetes, etc.

D. Blood is removed from the body, cleaned and returned

E. Inherited disorder in which masses form and destroy kidney tissue

F. Uses high energy sound waves to break up kidney stones

G. Deposits of uric acid, calcium salts and other substances in the renal pelvis

Short Answer

45. Why are females more prone to urinary tract infections than are males? [p.236] _____

_____

_____

_____

Fill-in-the-Blanks [p.236]

Kidneys can be damaged by over-the-counter (46) _____, (47) _____, and (48)

_____. (49) _____ provides a chemical snapshot of many processes in the body and can be

helpful in diagnosing disease. Glucose in the urine may be a sign of (50) _____. The presence of white

blood cells often indicates a(an) (51) _____ infection.

Short Answer

52. Explain how the urinary system contributes to homeostasis. [p.237] _____

_____

_____

_____

_____

____ 1. Which of these is not a route by which water leaves the body? [p.227]

    a. evaporation from lungs and skin
    b. sweating
    c. urine
    d. hair follicles
    e. feces

____ 2. The functional subunit of a kidney that filters blood, and restores solute and water balance is a slender tube called a _____. [p.228]

    a. renal capsule
    b. renal medulla
    c. nephron
    d. ureter
    e. none of the above

____ 3. In humans, the thirst center is located in the _____. [p.233]

    a. adrenal cortex
    b. thymus
    c. heart
    d. adrenal medulla
    e. brain

____ 4. Of all the water and sodium that leaves the bloodstream at the glomerulus, about two-thirds is promptly reabsorbed across the _____. [p.232]

    a. loop of Henle
    b. proximal tubule
    c. ureter
    d. Bowman's capsule
    e. collecting duct

____ 5. Along the ascending loop of Henle, sodium is moved out by _____. [p.232]

    a. phagocytosis
    b. countercurrent multiplication
    c. bulk flow
    d. active transport
    e. diffusion

____ 6. Filtration of the blood in the kidney takes place across the _____. [p.230]

    a. loop of Henle
    b. proximal tubule
    c. distal tubule
    d. glomerular capillaries
    e. all of the above

____ 7. _____ controls the reabsorption of sodium in the distal tubules and collecting ducts. [p.233]

    a. Insulin
    b. Glucagon
    c. Antidiuretic hormone
    d. Aldosterone
    e. Epinephrine

____ 8. _____ controls the reabsorption of water in the distal tubules and collecting ducts. [p.232]

    a. Insulin
    b. Glucagon
    c. Antidiuretic hormone
    d. Aldosterone
    e. Epinephrine

____ 9. The last portion of the excretory system that urine passes through before it is eliminated from the body is the _____. [p.228]

    a. renal pelvis
    b. bladder
    c. ureter
    d. collecting ducts
    e. urethra

____ 10. _____ is the principal waste product of protein breakdown, a combination of ammonia and carbon dioxide produced by the liver. [p.227]

    a. Urea
    b. Uric acid
    c. Creatine
    d. Amino acid
    e. ADH

CHAPTER OBJECTIVES/REVIEW QUESTIONS

1. List some of the factors that can change the composition and volume of body fluids. [pp.226-227]

2. List three electrolytes that the kidneys help to maintain correct levels of in the body. [p.227]

3. List successively the parts of the human urinary system that constitute the path of urine formation and excretion. [pp.228-229]

4. Locate the processes of filtration, reabsorption, and secretion along a nephron, and tell what makes each process happen. [pp.230-231]

5. Discuss the homeostatic role played by the kidneys in regulating fluid balance in the body. [p.232]

6. Explain the involvement of hormones in the function of the kidneys. [p.232-233]

7. Describe the role of the kidneys in maintaining the pH of the extracellular fluids between 7.37 and 7.43. [p.234]

8. Discuss the basis of urinalysis as a tool for diagnosing disease. [p.236]

9. Explain how kidney stones form and how they are treated. [p.235]

10. List three kidney disorders and explain what can be done if kidneys become too diseased to work properly. [p.235]

INTEGRATING AND APPLYING KEY CONCEPTS

1. The kidneys are commonly thought of as liquid-waste processing plants only. Consider the complex homeostatic interactions between the urinary system and other organ systems of the body. How might renal failure affect the other organ systems as well as the urinary system?

2. Currently, those in renal failure rely on dialysis or a kidney transplant to survive. Research the future medical hopes for dealing with insufficient kidney function.

13

THE NERVOUS SYSTEM

INTRODUCTION

Everything that you do while studying this chapter, from opening the textbook to thinking about the facts therein to daydreaming, is under the control of your nervous system. To begin to understand this complex body system, first master the series of events by which a message is carried along a nerve cell. Once you are confident with this information, then consider the functions of various parts of the nervous system, as well as how the brain controls it all.

FOCAL POINTS

- Figure 13.2 [p.242] illustrates the structure of a motor neuron. Since structure and function always reflect each other, consider how this specialized cell accomplishes its job within the nervous system.
- Figure 13.4 [pp.244-245] presents the chemical events involved in an action potential. This membrane transport process of creating gradients as a source of energy on demand is seen repeatedly in biological systems. The functioning of a nerve is just one example. This is also a very good example of a positive feedback event.
- Figure 13.9 [p.249] is an excellent visual showing how sensory and motor neurons are involved in a reflex. Following the lettered events of this diagram will help to simplify how nerves work.
- Figure 13.10 [p.250] visually summarizes the main parts of the nervous system.
- Figure 13.16 [p.254] delineates the three anatomical divisions of the human brain and identifies the parts of the brain that are included in these divisions.

INTERACTIVE EXERCISES

CHAPTER INTRODUCTION [p. 241]

13.1. NEURONS: THE COMMUNICATION SPECIALISTS [p.242-243]

13.2. NERVE IMPULSES = ACTION PTOENTIALS [pp.244-245]

13.3. HOW NEURONS COMMUNICATE [pp.246-247]

Boldfaced Terms

nervous system _____

_____

sensory neurons _____

_____

interneurons _____ _____

_____

motor neurons _____

_____

dendrites _____

_____

axon _____

_____

resting membrane potential _____

_____

threshold _____

_____

action potential _____

_____

sodium–potassium pumps _____

_____

neurotransmitters _____

_____

chemical synapse _____

_____

neuromodulators _____

_____

synaptic integration _____

_____

Matching

Write the letter of the matching definition by the terms below. [pp. 242-245]

1. _____ threshold level
2. _____ cell body
3. _____ action potential
4. _____ interneurons
5. _____ sodium-potassium pump
6. _____ sensory neurons
7. _____ neuroglia
8. _____ resting membrane potential
9. _____ dendrite
10. _____ motor neurons

A. Physically support and protect neurons; help maintain proper ion concentrations

B. A "nerve impulse"

C. Carrier protein; moves Na^+ and K^+ across membrane

D. Collect information about stimuli and relay it to the brain

E. Minimum shift in voltage difference required for an action potential

F. Steady charge difference across neuron cell membrane

G. Receive and process sensory input and send signals to other neurons

H. Neuron's "input zone"; receives incoming signals

I. Part of the neuron that contains the nucleus and organelles

J. Relay signals from interneurons to muscles and glands (effectors)

Short Answer

11. What are neuroglia and how do they work with neurons? [p.242] _____

_____

_____

_____

Labeling [p.242]

12.  _____

13. _____

14. _____

15. _____

16.  _____

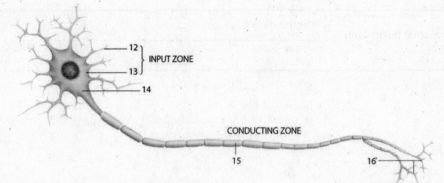

Dichotomous Choice

Circle one of the two possible answers given between parentheses in each statement. [pp.242-244]

17. Ions like Na^+ and K^+ can cross the nerve cell membrane through (the plasma membrane / channel proteins).

18. Shifts in the (charges / concentrations) of sodium and potassium set the stage for nervous system signals.

19. At rest, a neuron's gated sodium channels are (open / closed).

20. Of sodium and potassium ions, the neuron's plasma membrane allows more (sodium / potassium) to leak through.

21. Based on the rules of diffusion, sodium tends to move (in / out) and potassium tends to move (in / out).

22. Ion concentrations result in the cytoplasm just inside the membrane being (positively / negatively) charged compared to the fluid just outside the membrane.

23. The steady difference in charge across the membrane of −70 millivolts is called the resting membrane (differential / potential), indicating that the difference has the ability to do work.

24. A strong enough signal arrives, it causes the voltage difference to (increase / reverse).

25. A strong enough signal causes sodium gates to (open / close).

26. The threshold level of stimulation for a neuron will result in a (positive / negative) feedback situation in which more and more sodium channels open in response to previous ones.

27. A nerve impulse is correctly called a(n) (threshold stimulus / action potential).

28. For an action potential to occur, a stimulus must be strong enough to spread to the (input / trigger) zone.

29. To transmit messages throughout the body, an action potential must spread to other neurons or to cells in a muscle or (gland / sensory receptor).

True-False

If the statement is true, write a T in the blank. If the statement is false, make it correct by writing the word(s) in the blank that should take the place of the underlined word(s). [pp.244-245]

30. _____ Self propagation of an action potential can occur because the changes in membrane potential leading up to it <u>don't</u> lose strength.

31. _____ Action potentials always propagate <u>toward</u> a trigger zone.

32. _____ A neuron can't respond to an incoming signal unless concentration and electric <u>gradients</u> are in place across its membranes.

33. _____ A resting neuron uses energy to power a <u>facilitated diffusion</u> mechanism to maintain sodium and potassium gradients.

34. _____ Sodium-potassium pumps move potassium <u>out</u> of the cell and sodium in the opposite direction.

35. _____ Every action potential spikes to the same level as an <u>all-or-nothing</u> event.

36. _____ When a spike ends, the gated sodium channels close, while <u>chloride</u> channels are open so that ions flow out and restore the original voltage difference across the membrane.

Labeling

Label the parts of the accompanying illustration. [p.245]

37. _____

38. _____

39. _____

Time (milliseconds)

Fill-in-the-Blanks [pp. 245-247]

The arrival of a signal may prompt the release of one or more chemical messengers called (40)

_____. These molecules diffuse across a(n) (41) _____, a gap between the neuron's output

zone and the input zone of the neighboring neuron. The (42) _____ cell stores neurotransmitter

molecules. When an action potential arrives, gated channels open and (43) _____ ions flow from

outside the cell down their gradient into the cell. This results in synaptic (44) _____ fusing with the

plasma membrane and releasing neurotransmitters. These diffuse across the synapse and bind with receptor proteins

on the receiving, or (45) _____, cell. This binding causes a membrane channel to open so that ions can

flow through and enter the receiving cell.

Some neurotransmitters are excitatory and help drive the membrane to a(n) (46) _____, while

others are (47) _____ signals. Examples of neurotransmitters include (48) _____ or ACh,

and serotonin. Neuromodulators can magnify or impede the effects of a(n) (49) _____. An example is

the group of natural painkillers called (50) _____. These inhibit nerves from releasing substance P,

which conveys information about (51) _____. In athletes who exercise beyond normal fatigue, (52)

_____ can produce a euphoric (53) _____.

At any moment, many signals are washing over the (54) _____ of a receiving neuron. All of

them are (55) _____ competing for control of the membrane potential at the trigger zone. The ones

called EPSPs (56) _____ the membrane while the ones called IPSPs (57) _____ the

membrane. Synaptic integration tallies up competing signals in a process called (58) _____. (59)

_____ occurs when neurotransmitter molecules from more than one presynaptic cell reach a neuron's

input zone at the same time.

The flow of signals through the nervous system depends on the rapid, controlled (60) _____ of neurotransmitters from synapses. Some diffuse out, others are cleaved by (61) _____ in the synapse such as acetylcholinesterase. Some are actively (62) _____ back into the presynaptic cells.

13.4. INFORMATION PATHWAYS [PP.248–249]

13.5. OVERVIEW OF THE NERVOUS SYSTEM [PP.250–251]

Boldfaced Terms

nerve _____

_____

nerve tracts _____

_____

myelin sheath _____

_____

Schwann cells _____

_____

reflex _____

_____

central nervous system _____

_____

peripheral nervous system _____

_____

ganglia _____

_____

Fill-in-the-Blanks [pp. 248-249]

A(n) (1) _____ consists of the long axons of sensory neurons, motor neurons, or both. Each axon has a(n) (2) _____ that speeds the rate at which the (3) _____ propagates. The sheath consists of glia called (4) _____ cells. An exposed (5) _____, or gap, separates each Schwann cell from the next one. Action potentials jump from node to node in what is called (6) _____ conduction. In a large, sheathed axon, action potentials (7) _____ at the rate of 120 meters per second!

The (8) _____ has no Schwann cells. There, glia called (9) _____ sheath myelinated axons.

Sensory and motor neurons take part in automatic responses called (10) _____. A reflex is a simple (11) _____ movement that occurs in response to a stimulus. In the simplest (12) _____, sensory neurons synapse directly on motor neurons. The stretch reflex (13) _____ a muscle involuntarily when gravity or some other load has caused the muscle to stretch. In most reflex pathways, the sensory neurons interact with several (14) _____, which excite or inhibit motor neurons. This allows a coordinated response.

Sensory nerves relay information into the (15) _____ where they synapse with neurons. The brain and spinal cord contain only (16) _____ which integrate the signals. Many interneurons synapse with (17) _____ neurons that carry messages away from the spinal cord and brain.

Blocks of hundreds or thousands of interneurons in the brain and spinal cord are parts of (18) _____ within which integration of signals occurs. Some circuits fan out to many others or (19) _____, while others funnel many signals to a few. Others repeat signals among themselves and are called (20) "_____" circuits.

Labeling
Label the parts of the illustrated nerve. [p.248]

21. _____

22. _____

23. _____

24. _____

25. _____

Outer connective tissue

21
22
23
24
25

Matching

Match the following choices with the correct number in the diagram. [p.249]

26. ____

27. ____

28. ____

29. ____

30. ____

31. ____

32. ____

A. Muscle cell plasma membrane stimulated to contract (response)
B. Action potentials generated in motor neuron propagate along its axon toward muscle
C. Axon endings of motor neuron synapse with muscle cells, release neurotransmitter
D. Spinal cord
E. Receptor endings of stimulated sensory neuron generate action potential toward spinal cord
F. Muscle spindle
G. Axon endings of sensory neuron release neurotransmitter that stimulates motor neuron

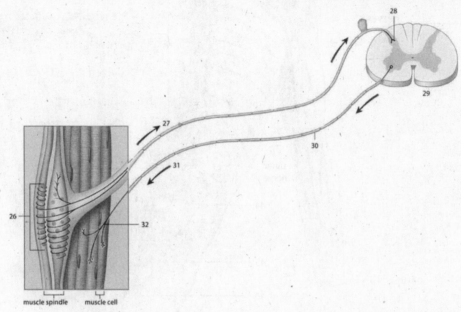

Fill-in-the-Blanks [p. 250]

The nervous system has two main regions. The brain and spinal cord, including all interneurons, comprise the (33) _____ or CNS. The motor and sensory nerves throughout the rest of the body are the (34) _____ or PNS. The PNS is organized into (35) _____ and (36) _____ subdivisions, with the latter being subdivided again. The PNS has 31 pairs of (37) _____ nerves, and 12 pairs of (38) _____ nerves. Places where the cell bodies of several neurons cluster together are called (39) _____.

Labeling

Label each numbered part of the accompanying illustration. [p.250]

40. _____ nerves

41. _____

42. _____ nerves

43. _____ nerves

44. _____ nerves

45. _____ nerves

46. _____ nerves

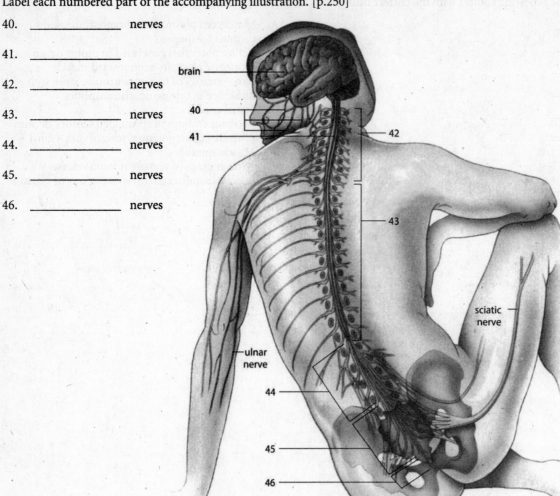

brain

40

41

42

43

sciatic
nerve

ulnar
nerve

44

45

46

On the blank line next to the description of function, place the Roman numeral of the appropriate cranial nerve. [p.251]

47. ____ Olfactory nerve

48. ____ Optic nerve (from the retina)

49. ____ From inner ear

50. ____ To jaw muscles; from mouth

51. ____ To facial muscles, glands; from taste buds

13.6. MAJOR EXPRESSWAYS: PERIPHERAL NERVES AND THE SPINAL CORD [PP.252–253]

13.7. THE BRAIN: COMMAND CENTRAL [PP.254–255]

13.8. A CLOSER LOOK AT THE CEREBRUM [PP.256–257]

Boldfaced Terms

somatic nerves _____

_____

autonomic nerves _____

_____

parasympathetic nerves _____

_____

sympathetic nerves _____

_____

spinal cord _____

_____

white matter _____

_____

gray matter _____

_____

brain _____

_____

medulla oblongata _____

_____

meninges _____

_____

brain stem _____

_____

hindbrain _____

_____

pons _____

_____

cerebellum _____

_____

brain stem_____

_____

cerebrum _____

_____

cerebral hemispheres _____

_____

thalamus _____

_____

hypothalamus _____

_____

meninges _____

_____

cerebrospinal fluid _____

_____

blood-brain barrier _____

_____

cerebral cortex _____

_____

motor areas _____

_____

sensory areas _____

_____

association areas _____

_____

limbic system _____

_____

Choice

Choose A or S to match each statement about the peripheral nervous system. [p. 252]

      A. autonomic nerves   S. somatic nerves

1. _____ Signals travel to and from internal organs and other structures

2. _____ Sensory axons carry information from receptors in skin, skeletal muscles, and tendons

3. _____ Signals concern moving the head, trunk, and limbs

4. _____ Includes preganglionic and postganglionic neurons

5. _____ Motor axons carry messages to smooth muscle, cardiac muscle, and glands

6. _____ Motor axons deliver commands to skeletal muscles

7. _____ Includes parasympathetic and sympathetic nerves

Dichotomous Choice

Circle one of the two possible answers given between parentheses in each statement. [pp. 252-253]

8. (Sympathetic / Parasympathetic) nerves cause the pupils to constrict and heart rate to decrease, as well as increasing stomach and intestinal movements.

9. (Sympathetic / Parasympathetic) nerves cause glandular secretions in the airways to decrease and salivary gland secretions to thicken.

10. (Sympathetic / Parasympathetic) nerves tend to slow down the body when there is not much outside stimulation.

11. (Sympathetic / Parasympathetic) nerves tend to speeds up the body during heightened awareness, excitement, or danger.

12. (Sympathetic/Parasympathetic) nerves inhibit glucose release from the liver.

13. The (gray matter / white matter) is found on the inside of the spinal cord.

14. The (gray matter / white matter) contains dendrites, cell bodies of neurons, interneurons, and neuroglial cells.

15. The spinal cord is protected by the vertebral column, as well as protective coverings called the (intervertebral disks / meninges).

16. (Autonomic reflexes/Spinal reflexes) deal with internal functions such as bladder emptying.

17. The spinal cord carries signals between the peripheral nerves and (the brain/the organs of the body).

Labeling

Identify the numbered parts of the accompanying illustration. [p.253]

18. _____   22. _____

19. _____   23. _____

20. _____   24. _____

21. _____   25. _____

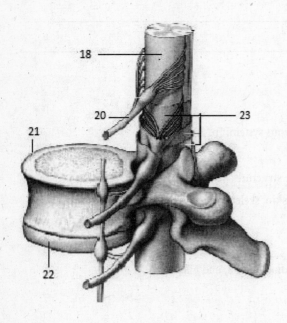
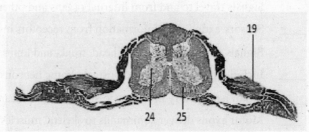

Fill-in-the-Blanks [p. 254]

The spinal cord merges with the (26) _____, a master control center. The brain is protected by the bones of the (27) _____ and by the three (28) _____. Folds in the tough, outermost (29) _____ mater separate the brain into right and left (30) _____. A thinner middle layer is called the (31) _____ mater and an even more delicate (32) _____ mater wraps the brain and spinal cord. The meninges also enclose spaces filled with fluid that (33) _____ and helps nourish the brain. Information is processed and sensory input and motor responses are integrated in the (34) _____. The structure that connects the two hemispheres is the (35) _____ .

Matching

To the right of each number, indicate which division of the brain includes the structure by writing H for hindbrain, M for midbrain, or F for forebrain. Then, in the parentheses, match the named part with the letter of the phrase that describes its function. [p.254]

36. ____ ( ) cerebrum

37. ____ ( ) hypothalamus

38. ____ ( ) cerebellum

39. ____ ( ) medulla oblongata

40. ____ ( ) pons

41. ____ ( ) thalamus

A. Monitors internal organs; influences behaviors related to thirst, hunger, sexual behavior and emotional expression

B. Information is processed, and sensory input and motor responses are integrated

C. Relays and coordinates sensory signals to cerebrum through clustered neuron cell bodies (nuclei)

D. Directs signal traffic between cerebellum and forebrain

E. Coordinates sensory and motor nerve signals for movement and balance

F. Site of reflex centers involved in respiration and blood circulation

Fill-in-the-Blanks [pp. 254-255]

The brain and spinal cord are surrounded by transparent (42) _____ fluid. It is secreted from

specialized capillaries inside cavities called (43) _____. The fluid is also found in the central canal of the

(44) _____ and the brain itself, helping to cushion the brain and spinal cord from jarring movements.

The structure of brain capillaries forces substances to pass through endothelial cells, not between them, in order to

reach the brain. This (45) _____ helps control which substances reach the brain's neurons. The

"loophole" in the system allows (46) _____ -soluble substances through, including caffeine, nicotine,

alcohol, and many illegal drugs.

True/False

If the statement is true, write a T in the blank. If the statement is false, make it correct by writing the word(s) in the blank that should take the place of the underlined word(s). [254-257]

47. _____ The <u>cerebellum</u> is divided into right and left hemispheres.

48. _____ The cerebral cortex is a thin, outer layer of <u>gray</u> matter.

49. _____ The <u>left</u> hemisphere deals with visual-spatial relationships, music and other creative activities.

50. _____ The <u>left</u> hemisphere dominates the right hemisphere in most people.

51. _____ The corpus callosum is a band of <u>connective tissue</u> between the hemispheres.

52. _____ Each hemisphere is divided into four regions called <u>lobes</u>.

53. _____ Everything people comprehend, communicate, remember, and voluntarily act upon arises in the <u>brain stem</u>.

54. _____ Motor areas in the primary motor cortex of the frontal lobe control coordinated movements of skeletal muscles.

55. _____ The premotor cortex deals with instinctive behaviors.

56. _____ Broca's area used in speech, as well as the eye field controlling voluntary eye movements, is usually in the left hemisphere.

57. _____ The main receiving center for sensory input from the skin and joints is in the frontal lobe.

58. _____ Taste is perceived in the parietal lobe, sight in the occipital lobe, and sound and smell in the temporal lobes.

59. _____ Association areas in all parts of the cortex integrate, analyze, and respond to many inputs.

60. _____ The limbic system is located on the outside of the cerebral hemispheres.

61. _____ The limbic system governs emotions and has roles in memory

62. _____ The arachnoid wraps the brain and spinal cord.

63. _____ The blood-brain barrier does not protect the hypothalamus.

64. _____ Thumb, finger and tongue muscles get much attention from Broca's area.

Labeling

Identify each numbered part of the accompanying illustration. [p.254]

65. _____

66. _____

67. _____

68. _____

69. _____

70. _____

71. _____

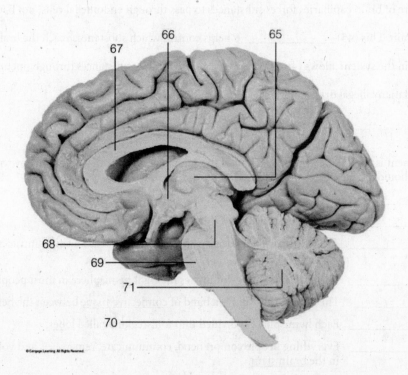

13.9. CONSCIOUSNESS [P.258]

13.10. MEMORY [P.259]

13.11. DISORDERS OF THE NERVOUS SYSTEM [PP.260-261]

13.12. THE BRAIN ON "MIND-ALTERING" DRUGS [P.262]

13.13. THE NERVOUS SYSTEM IN HOMEOSTASIS [P.263]

Boldfaced Terms

reticular formation _____

_____

memory _____

_____

amnesia _____

_____

concussion _____

_____

paralysis _____

_____

seizure disorders _____

_____

Parkinson's disease _____

_____

Alzheimer's disease _____

_____

meningitis _____

_____

encephalitis _____

_____

Creuzfeldt-Jakob disease _____

_____

glial cancers _____

_____

multiple sclerosis (MS) _____

_____

Guillain-Barre syndrome _____

_____

headache _____

_____

ADHD _____

_____

mood disorders _____

_____

autism _____

_____

schizophrenia _____

_____

Matching

Match the disease or condition to its correct description. [pp. 260-261]

1. _____ multiple sclerosis

2. _____ epilepsy (seizure disorders)

3. _____ meningitis

4. _____ concussion

5. _____ Creuzfeldt-Jakob disease

6. _____ encephalitis

7. _____ Parkinson's disease

8. _____ Alzheimer's disease

9. _____ migraine

A. Slow death of basal nuclei of thalamus; produces muscle tremors and balance problems

B. Inflammation of the brain usually due to viral infection

C. Autoimmune disease causing destruction of the myelin sheaths of the central nervous system

D. Brain's normal electric activity becomes chaotic

E. Results from violent blow to head or neck

F. Degeneration of brain neurons and buildup of amyloid protein, leading to loss of memory and intellect

G. Throbbing headache accompanied by nausea, vomiting and light sensitivity

H. Often fatal inflammation of the meninges caused by viral or bacterial infection

I. Develops after eating beef from animals infected by a prion

Matching

Match the chemical substance to its correct functional description. [p.262]

10. ____ amphetamine

11. ____ nicotine

12. ____ cocaine

13. ____ alcohol

14. ____ morphine

15. ____ OxyContin

16. ____ marijuana

A. Diminishes judgment and may produce disorientation and lack of coordination; depresses brain activity

B. Slows motor activity, elicits a mild euphoria, skews performance of complex tasks

C. Causes a flood of dopamine and norepinephrine, thus stimulating the brain's pleasure center

D. Synthetic prescription drug version of morphine, produces euphoria

E. Mimics acetylcholine by stimulating sensory receptors, increases heart rate and blood pressure

F. Blocks pain signals by binding with certain CNS receptors; from opium poppy seed pods; similar to heroin

G. Blocks reabsorption of dopamine and other neurotransmitters, thus stimulating pleasure centers

Short Answer [p.262]

17. Tolerance develops when detoxifying liver enzymes increase to the point that the entire drug in the bloodstream is broken down before it can have an effect. What must an addict do to get past tolerance? _____

_____

_____

_____

18. What is habituation? _____

_____

_____

Matching

Match the term to its correct description. [pp. 256, 258]

19. _____ reticular formation

20. _____ slow-wave sleep

21. _____ EEG

22. _____ PET

23. _____ REM sleep

A. Shows exact location of brain activity as it takes place

B. A two-part network of interconnected neurons that runs through the brain stem

C. Recording of the summed electrical activity of the brain's neurons

D. Dreaming occurs, eyelids flicker and eyeballs move

E. "Normal" sleep in which heart rate, breathing and muscle tone change very little

SELF-QUIZ

_____ 1. The conducting zone of a neuron is the _____. [p.242]

    a. axon
    b. axon endings
    c. cell body
    d. dendrite
    e. axon hillock

_____ 2. An action potential is brought about by _____. [p.244]

    a. a sudden membrane impermeability
    b. the movement of negatively charged proteins through the neuronal membrane
    c. the movement of lipoproteins to the outer membrane
    d. a local change in membrane permeability caused by a greater-than-threshold stimulus
    e. an order sent from the neuron's DNA

_____ 3. The resting membrane potential _____. [p.243]

    a. exists as long as a charge difference that could do physiological work exists across a membrane
    b. occurs because there are more potassium ions outside the neuronal membrane than there are inside
    c. occurs because of the unique distribution of receptor proteins located on the dendrite exterior
    d. is brought about by a local change in membrane permeability caused by a greater-than-threshold stimulus
    e. occurs only when a person is resting or sleeping.

_____ 4. The phrase "all-or-nothing" used in conjunction with a discussion about an action potential means that _____. [p.245]

    a. a resting membrane potential has been received by the cell
    b. it will always spike totally once stimulated past threshold
    c. the membrane either achieves total equilibrium or remains as far from equilibrium as possible
    d. propagation along the neuron is saltatory
    e. all of the Na^+ and K^+ along a neuron's membrane switch places

_____ 5. _____ are responsible for integration of sensory and motor signals within the central nervous system. [p.242]

    a. Interneurons
    b. Schwann cells
    c. Motor neurons
    d. Sensory neurons
    e. Neuroglia

_____ 6. _____ nerves generally dominate during quiet, low-stress situations. [p.252]

    a. Ganglia
    b. Pacemaker
    c. Sympathetic
    d. Parasympathetic
    e. Somatic

____ 7. The center of consciousness, memory, and intelligence is the _____. [p.254]

    a. hindbrain
    b. reticular formation
    c. midbrain
    d. brain stem
    e. forebrain

____ 8. The left cerebral hemisphere is generally responsible for _____. [p.256]

    a. music
    b. mathematics
    c. spatial relationships
    d. abstract abilities
    e. artistic ability

____ 9. The part of the brain that controls the basic responses necessary to maintain life processes (homeostatic adjustments) is the _____. [p.254]

    a. cerebral cortex
    b. cerebellum
    c. corpus callosum
    d. medulla oblongata
    e. hypothalamus

____ 10. The center for coordination of voluntary movements is the _____. [p.254]

    a. cerebrum
    b. pons
    c. cerebellum
    d. hypothalamus
    e. thalamus

CHAPTER OBJECTIVES/REVIEW QUESTIONS

1. Draw a neuron and label it according to its three general zones, its specific structures, and the specific function(s) of each structure. [p.242]

2. Describe the process that leads to the creation of an action potential. [p.244]

3. Explain what a reflex is by drawing and labeling a diagram and telling how it functions. [p.249]

4. Define and contrast the central and peripheral nervous systems. [pp.250-251]

5. Distinguish between somatic and autonomic nerves. [p.250]

6. Explain how parasympathetic nerve activity balances sympathetic nerve activity. List activities of the sympathetic and parasympathetic nerves in regulating pupil diameter, rate of heartbeat, activities of the gut, and action of sphincter muscles. [p.252]

7. Describe the structure and function of the blood-brain barrier. [p.255]

8. List the parts of the brain found in the hindbrain, midbrain, and forebrain, and tell the basic functions of each. [pp.254-255]

9. Define motor, sensory and association areas and list the functions of each. [p.256-257]

10. Explain why the limbic system is called our "emotional brain." [p257]

11. Define and discuss the characteristics of slow-wave sleep and REM sleep. [p.258]

12. Define amnesia and discuss the factors that determine its severity. [p.259]

13. Describe six or more major disorders/diseases of the human nervous system by naming the causes and symptoms of each. [pp.260-261]

14. Name five drugs that are commonly abused and state their effects on the brain. [p.262]

15. Discuss how the nervous system acts homeostatically with the skeletal system, immune system and endocrine system. [p.263]

INTEGRATING AND APPLYING KEY CONCEPTS

Suppose that anger is eventually determined to be caused by excessive amounts of specific transmitter substances in the brains of angry people. Also suppose that an inexpensive antidote that neutralizes these anger-producing transmitter substances is readily available. Can violent murderers now argue that they have been wrongfully punished because they were victimized by their brain's transmitter substances and could not have acted in any other way? Suppose an antidote is prescribed to curb violent temper in an easily angered person. Suppose also that the person forgets to take the pill and subsequently murders a family member. Can the murderer still claim to be victimized by transmitter substances?

14

SENSORY SYSTEMS

INTRODUCTION

This chapter is an extension of the previous chapter covering the nervous system. It focuses on the senses, specialized organs that allow us to connect with, and interpret, our environment. This chapter examines the many different types of sensory receptors, the stimuli that they receive and the pathways to the areas of the brain that interpret them. This chapter will provide explanations for what allows us to see, hear, taste, smell and feel.

FOCAL POINTS

- Table 14.1 [p.268] provides a good summary of types of sensory receptors and their individual stimuli
- Figure 14.1 [p.269] diagrams a sensory pathway
- Figure 14.2, animated, [p.270] provides a map of the somatosensory cortex
- Figure 14.3, animated, [p.271] diagrams the various sensory receptors in the skin
- Figure 14.4, animated, [p.271] indicates the surface manifestations of referred pain
- Figure 14.5, animated, [p.272] maps the location of taste receptors on the tongue
- Figure 14.6, animated, [p.273] outlines the olfactory sensory pathway
- Figure 14.7, animated, [p.274] demonstrates how sound waves vary in pitch and frequency
- Figure 14.8, animated, [pp.274-275] illustrates the structure of the auditory apparatus
- Figures 14.9, animated, and 14.10 [p.276] illustrate the structures of the ear responsible for equilibrium
- Table 14.2 [p.278] provides a good summary of the structures of the eye and their functions
- Figure 14.13, animated, [p.278] illustrates the structures of the eye
- Figure 14.15, animated, [p.279] demonstrates the accommodation of the lens of the eye
- Figure 14.19, animated, [p.281] diagrams the sensory pathway of sight
- Figure 14.20, animated, [p.282] explains nearsightedness and farsightedness

INTERACTIVE EXERCISES

CHAPTER INTRODUCTION [p. 267]

14.1. SENSORY RECEPTORS AND PATHWAYS [pp.268-269]

14.2. SOMATIC SENSATIONS [pp.270-271]

Boldfaced Terms

stimulus _____

_____

sensation _____

_____

perception _____

_____

mechanoreceptors _____

_____

thermoreceptors _____

_____

nociceptors _____

_____

chemoreceptors _____

_____

osmoreceptors _____

_____

photoreceptors _____

_____

sensory adaptation _____

_____

somatic sensations _____

_____

somatosensory cortex _____

_____

free nerve endings _____

_____

encapsulated receptors _____

_____

Short Answer [268-269]

1. Distinguish between a sensation and a perception. _____

_____

_____

2. List the three ways in which the brain determines the nature of a given stimulus. _____

_____

_____

3. Name one sensation with which sensory adaptation occurs. _____

_____

_____

_____

4. Distinguish between somatic sensations and special senses. _____

_____

_____

_____

Choice

Choose the type of sensory receptor associated with each phrase. [p.268]

    a. chemoreceptors   b. mechanoreceptors   c. nociceptors   d. photoreceptors

              e. thermoreceptors   f. osmoreceptors

5. _____ Detect taste of substances dissolved in saliva

6. _____ Detect visible light

7. _____ Stimulated by tissue damage

8. _____ Olfactory receptors in nose are an example

9. _____ Include auditory receptors

10. _____ Detect CO_2 concentration in extracellular fluid

11. _____ Detect temperature change

12. _____ Stimulated by change in water volume of surrounding fluid

13. _____ Stimulated by mechanical pressure against body

14. _____ Rods and cones are examples

15. _____ Includes baroreceptors and balance

16. ____ Located in hypothalamus

17. ____ Stimulated by stretching

Dichotomous Choice

Circle one of two possible answers given between parentheses in each statement. [pp. 268-271]

18. Somatic sensations travel from body receptors to the spinal cord, and then to the somatosensory cortex in the brain's (cerebellum / cerebrum).

19. Interneurons in the somatosensory cortex form a "map" of the body's surface with the largest areas corresponding to the body parts where the density of sensory receptors is (least / greatest).

20. You might feel a spider walking on your arm because of (chemoreceptors / mechanoreceptors).

21. Encapsulated receptors that respond to light touching are (Meissner's corpuscles / free nerve endings).

22. Receptors that respond to steady touching and pressure, as well as to high temperature, are called (Pacinian corpuscle / Ruffini endings).

23. Pain associated with internal organs is called (somatic pain / visceral pain).

24. Damaged cells release chemicals such as bradykinins that (cause immediate pain / activate pain receptors).

25. Sensing the pain of a heart attack along the left shoulder is an example of (referred pain / phantom pain).

14.3. TASTE AND SMELL: CHEMICAL SENSES [pp.272-273]

14.4. TASTY SCIENCE [p.273]

Boldfaced Terms

chemical senses _____

_____

taste receptors _____

_____

olfactory receptors _____

_____

Fill-in-the-Blanks [272-273]

In both taste and smell, (1) _____ bind molecules that are dissolved in the fluid bathing them.

Sensory information travels from the receptors through the (2) _____ and on to the cerebral cortex. In

the case of taste, these receptors are often part of sense organs called (3) _____ such as those scattered

over the tongue. These distinguish only (4) _____ basic types of flavors: sweet, sour, (5)

_____, bitter, and (6) _____.

Animals smell substances by means of (7) _____ receptors that detect water-soluble or volatile substances. Sensory nerve pathways lead from the nasal cavity to the region of the brain called the (8) _____. From there, other neurons forward the message to a center in the (9) _____ , which interprets it as a particular smell. About half an inch inside the nose is the (10) _____ organ, or "sexual nose." It detects signal molecules called (11) _____. These affect the behavior of other individuals.

Each of the five categories of taste is associated with particular (12) _____ molecules. Each taste bud has receptors that can respond to tastants in at least (13) _____ and sometimes all five of the taste classes. Some taste receptors, such as "bitter" are extremely (14) _____, while others require much more stimulation in order to release a neurotransmitter.

14.5. HEARING: DETECTING SOUND WAVES [pp.274-275]

14.6. BALANCE: SENSING THE BODY'S NATURAL POSITION [pp.276-277]

14.7. DISORDERS OF THE EAR [p.277]

Boldfaced Terms

cochlea _____

_____

tympanic membrane _____

_____

organ of Corti _____

_____

hair cells _____

_____

tectorial membrane _____

_____

vestibular apparatus _____

_____

semicircular canals _____

_____

otitis media _____

_____

tinnitus _____

_____

deafness _____

_____

Fill-in-the-Blanks [p. 274]

Sounds are waves of compressed (1) _____. They are a form of (2) _____ energy.

The loudness of a sound corresponds to the (3) _____ of its wave form. The number of wave cycles per

second is the sound's (4) _____. The sense of hearing starts with vibration-sensitive (5)

_____ deep in the ear. When sound waves travel down the (6) _____, they reach a (7)

_____ and make it vibrate. These vibrations cause a fluid inside the ear to move. The moving fluid bends

the tips of (8) _____ on the mechanoreceptors. Enough bending results in a(n) (9) _____

that ultimately is interpreted by the brain as (10) _____.

Choice

Choose the region of the ear in which you will find each of the following. [p.274]

a. outer ear   b. middle ear   c. inner ear

11. _____ eardrum

12. _____ cochlea

13. _____ auditory canal

14. _____ semicircular canals

15. _____ pinna

16. _____ malleus, incus, stapes

Sequencing

In the blanks beside the numbers, write the letters of the ear parts to show the pathway of sound vibrations through the ear. [pp.274-275]

17. ____

18. ____

19. ____

20. ____

21. ____

22. ____

23. ____

24. ____

25. ____

26. ____

A. round window

B. stapes

C. fluid of cochlear duct

D. malleus

E. organ of Corti hair cells pushing against tectorial membrane (trigger action potential)

F. oval window

G. incus

H. tympanic membrane

I. auditory canal

J. fluid inside scala vestibuli and scala tympani

Labeling

Identify each indicated part of the accompanying illustration. [p.274]

27. _____

28. _____

29. _____

30. _____

31. _____

32. _____

33. _____

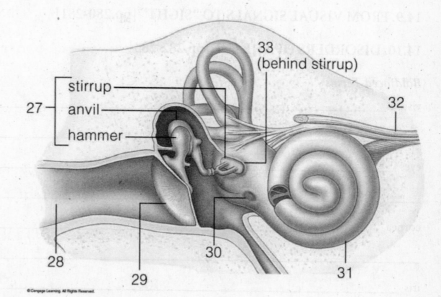

Matching

Select the description that best matches the numbered term. [pp. 275-277]

34. ____ tectorial membrane

35. ____ otitis media

36. ____ tinnitus

37. ____ decibels

38. ____ motion sickness

39. ____ otoliths

40. ____ semicircular canals

41. ____ eustachian tube

42. ____ 130 decibels

A. Equalizes pressure in the middle ear with the outside air

B. "Ear stones" that signal changes in the head's orientation relative to gravity

C. A ringing, whirring or buzzing in the ears

D. Occurs when extreme or continuous movement overstimulates hair cells in the balance organs

E. A measurement of the loudness of sounds

F. Positioned at right angles to each other corresponding to the three planes of space

G. Permanent damage may quickly occur to the sensory hair cells in the inner ear

H. A painful inflammation of the middle ear usually caused by a respiratory infection spread through the eustachian tubes

I. Jellylike structure contacted by hair cells that convert vibrations to nerve impulses that travel to the brain for sound interpretation

14.8. VISION: AN OVERVIEW [pp.278-279]

14.9. FROM VISUAL SIGNALS TO "SIGHT" [pp.280-281]

14.10. DISORDERS OF THE EYE [pp.282-283]

Boldfaced Terms

vision _____

_____

eyes _____

_____

cornea _____

_____

iris _____

_____

lens _____

_____

retina _____

_____

visual cortex _____

_____

accommodation _____

_____

rod cells _____

_____

cone cells _____

_____

fovea _____

_____

red-green color blindness _____

_____

astigmatism _____

_____

myopia _____

_____

hyperopia _____

_____

conjunctivitis _____

_____

herpes infection _____

_____

malignant melanoma _____

_____

retinoblastoma _____

_____

cataracts _____

_____

macular degeneration _____

_____

glaucoma _____

_____

retinal detachment _____

_____

Fill-in-the-Blanks [p.278]

Vision requires a system of (1) _____, and the (2) _____that receives nerve

impulses from the optic nerve. The sense of (3) _____ is an awareness of the position, shape, brightness,

distance, and movement of visual stimuli. The (4) _____ are sensory organs that contain a tissue with a

dense array of photoreceptors. The outer layer of the eye consists of a sclera and transparent (5) _____.

The middle layer includes the choroid, ciliary body, and (6) _____. The (7) _____ in the

center of the iris allows the entrance of light. The (8) _____ focuses incoming light onto a layer of

photoreceptor cells in the (9) _____. Photoreceptors are linked with neurons that form the (10)

_____.

Matching [p.278]

Choose the most appropriate description for each term.

11. ____ pupil
12. ____ sclera
13. ____ visual cortex
14. ____ lens
15. ____ vitreous humor
16. ____ ciliary body
17. ____ retina
18. ____ aqueous humor
19. ____ choroid
20. ____ cornea
21. ____ iris

A. Part of brain where signals are interpreted as sight

B. Focuses incoming light onto the retina

C. Pigmented ring behind cornea, adjusts amount of light entering eye

D. Pigmented area beneath sclera, prevents light from scattering

E. Jellylike substance in chamber behind lens

F. Clear fluid bathing both sides of lens

G. Smooth muscle that adjusts the shape of the lens to focus light

H. "Hole" through which light enters eye

I. Transparent covering in front of the iris and pupil

J. Layer of neural tissue at back of eye

K. Fibrous "white" of eye

Labeling

Identify each indicated part of the accompanying illustration. [p.278]

22. _____
23. _____
24. _____
25. _____
26. _____
27. _____
28. _____
29. _____
30. _____
31. _____
32. _____
33. _____
34. _____

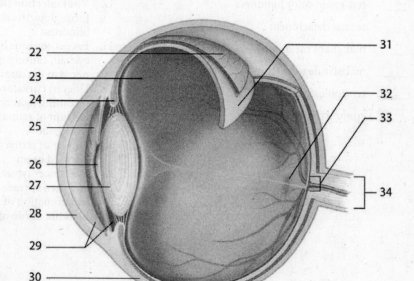

Dichotomous Choice

From the choice of terms in parentheses, select the term that best completes the statement. [280-281]

35. Daytime vision and color perception is the job of (rods / cones).

36. (Rods / cones) are better than (rods / cones) at detecting light intensity.

37. (Vitamin A / vitamin D) is used in making the pigment rhodopsin.

38. In the part of the retina called the (fovea / optic disk), visual acuity is higher than in the rest of the retina.

39. The optic nerves form from the axons of (bipolar neurons / ganglion cells).

40. In an early embryo, the (retinas / lenses) arise from the developing brain.

41. The part of the outside world that a person actually sees is his or her (receptive field / visual field).

42. The left and right visual fields of each eye (criss-cross / go straight) along the optic nerve to the brain.

Matching

Choose the most appropriate description for each term. [pp. 280-283]

43. ____ night blindness
44. ____ astigmatism
45. ____ cataracts
46. ____ hyperopia
47. ____ glaucoma
48. ____ herpes infection
49. ____ myopia
50. ____ red-green color blindness
51. ____ retinal detachment
52. ____ malignant melanoma
53. ____ macular degeneration
54. ____ retinoblastoma
55. ____ conjunctivitis

A. Caused by a physical blow to the head or an illness that separates the retina from the choroid
B. Caused by a deficiency in vitamin A needed to form visual pigments
C. Images of distant objects are focused in front of the retina; eyeball is wider than it is high
D. Gradual clouding of the lens
E. Inherited abnormality; retina lacks a particular type of cone cell
F. Close image are focused behind the retina; eyeball is taller than it is wide
G. Viral infection transmitted to baby when passing through birth canal of mother, may cause blindness
H. Excess aqueous humor accumulates inside the eyeball, causing neurons in the retina and optic nerve to die due to increased pressure
I. Uneven curvature of the cornea; cannot bend incoming light rays to the same focal point
J. Portion of retina breaks down and is replaced by scar tissue
K. Cancer of retina that can spread along optic nerve to brain
L. Eye cancer typically developing in the choroid and then spreads to other parts of the body
M. Inflammation of transparent membrane that lines the inside of eyelids and covers the sclera

SELF-QUIZ

____ 1. What type of sensory receptor senses limb motions and the body's position in space? [p.268]

a. mechanoreceptor
b. thermoreceptor
c. nociceptor
d. chemoreceptor

____ 2. What type of sensory receptor senses pain? [p.268]

a. mechanoreceptor
b. thermoreceptor
c. nociceptor
d. chemoreceptor

____ 3. A diminishing response to an ongoing stimulus is a (an) _____. [p.269]

a. perception
b. accommodation
c. sensory adaptation
d. somatic sensation

____ 4. Which type of pain is sensed by receptors in skin, skeletal muscles, joints and tendons? [p.270]

a. somatic pain
b. visceral pain
c. referred pain
d. phantom pain

____ 5. Olfactory bulbs receive sensory signals from _____. [p.272]

a. mechanoreceptors
b. thermoreceptors
c. nociceptors
d. chemoreceptors

____ 6. The principal place in the ear where sound waves are amplified is the _____. [p.274]

a. pinna
b. ear canal
c. oval window
d. organ of Corti

____ 7. In the ear, vibrations are translated into patterns of nerve impulses in the _____. [p.275]

a. pinna
b. ear canal
c. middle ear
d. organ of Corti

____ 8. Myopia (nearsightedness) is caused by _____. [p.282]

a. eye structure that focuses an image in front of the retina
b. uneven curvature of the lens
c. eye structure that focuses an image posterior to the retina
d. uneven curvature of the cornea

____ 9. Accommodation involves the ability to _____. [p.279]

a. change the sensitivity of the rods and cones by means of transmitters
b. change the width of the lens by relaxing or contracting certain muscles
c. change the curvature of the cornea
d. adapt to large changes in light intensity

____ 10. Rods and cones are photoreceptors found in the _____. [p.280]

a. iris
b. lens
c. choroid
d. retina

CHAPTER OBJECTIVES/REVIEW QUESTIONS

1. Define and distinguish among chemoreceptors, mechanoreceptors, photoreceptors, osmoreceptors and thermoreceptors. Name one example of each type that appears in humans. [p.268]

2. Describe the three steps in the processing of sensory information. [p.269]

3. Describe the function of nociceptors. [p.268]

4. Distinguish among somatic, visceral, referred and phantom pain. [pp.270-271]

5. What are the five primary tastes? [p.272]

6. Explain how a taste bud works. [p.272-273]

7. Follow a sound wave from the outer ear to the organ of Corti; mention the name of each structure it passes and state where the sound wave is amplified and where the pattern of pressure waves is translated into nervous impulses. [pp.274–275]

8. State how low- and high-frequency sounds affect the basilar membrane and the organ of Corti. [p.275]

9. Explain the roles of the oval and round windows in hearing. [pp.274-275]

10. Explain how the three semicircular canals of the human ear detect changes of position and acceleration in a variety of directions. [p.276]

11. Describe the structure of the human eye. [p.278]

12. Describe how the human eye perceives color and black-and-white. [p.280]

13. Explain the general principles that affect how light is detected by photoreceptors and changed into electrochemical messages. [p.280]

14. Define nearsightedness and farsightedness and relate each to eyeball structure. [p.282]

15. Describe and tell the cause of the following eye diseases: conjunctivitis, Herpes infection, malignant melanoma, retinoblastoma, cataracts, macular degeneration, glaucoma and retinal detachment. [pp.282-283]

INTEGRATING AND APPLYING KEY CONCEPTS

1. Discuss the benefits and problems associated with iris scans as a form of identification. Can you think of any other sensory structures or characteristics that are being, or could be, used for identification?

2. Explain how damage to the right side of the visual cortex might affect perception of light signals in the visual fields of both eyes.

15

THE ENDOCRINE SYSTEM

INTRODUCTION

This chapter begins with a discussion of hormones, what they are and what they do. Extremely small amounts of these chemicals have important effects on the body. The chapter continues by presenting the hormones and the components of the endocrine system that produce them. This information is vital to developing a valuable appreciation for the effects that stress and diet have on bodies.

FOCAL POINTS

* Figure 15.1 [p.289] shows the main components of the endocrine system. Studying this diagram along with Tables 15.2 [p.290] and 15.4 [p.294] will provide a general overview of the workings of the endocrine system.
* Figures 15.3 [p.292] and 15.4 [p.293] are examples of the complementary roles of the nervous and endocrine systems as agents of homeostasis. By reading chapters 13-15, you should to able to appreciate the interactions among all body systems.

INTERACTIVE EXERCISES

CHAPTER INTRODUCTION [P. 287]

15.1. THE ENDOCRINE SYSTEM: HORMONES [PP.288-289]

15.2. TYPES OF HORMONES AND THEIR SIGNALS [PP.290-291]

Boldfaced Terms

target cell _____

_____

hormones _____

_____

endocrine system _____

_____

opposing interaction _____

_____

synergistic interaction _____

permissive interaction _____

_____

steroid hormones _____

_____

nonsteroid hormones _____

_____

second messenger_____

_____

Matching

Choose the most appropriate description for each term. [p.288]

1. _____ hormones
2. _____ opposing interaction
3. _____ permissive interaction
4. _____ target cells
5. _____ synergistic interaction
6. _____ endocrine system
7. _____ prostaglandins

A. Sum total of the actions of two or more hormones necessary to produce the required effect on target

B. Group of glands, organs and cells that release hormones

C. Local signaling molecules that change conditions in nearby tissues

D. Effect of one hormone works against the effect of another

E. Secretions from endocrine glands or cells, and some neurons distributed by the bloodstream to target cells

F. Cells that have receptors for a given type of signaling molecule

G. One hormone exerts its effect only when a target cell has been "primed" to respond to that hormone

Complete the Table

8. Complete the following table by identifying the numbered components of the endocrine system shown in the illustration on the following page as well as the hormones produced by each gland. [p.289]

Gland Name	Number	Hormone(s) Secreted
a. parathyroids (four)		
b. adrenal gland cortex		1)
		2)
		3)
c. adrenal gland medulla		1)
		2)
d. pancreatic islets		1)
		2)
e. ovaries		1)
		2)
f. testes		
g. hypothalamus		1) *six* *(to ant. pituitary)*
		2) *(to post. pituitary)*
		3) *(to post. pituitary)*
h. pituitary, anterior lobe		1)
		2)
		3)
		4)
		5)
		6)
i. pituitary, posterior lobe		1)
		2)
j. thyroid		1)
		2)
k. thymus		
l. pineal gland		

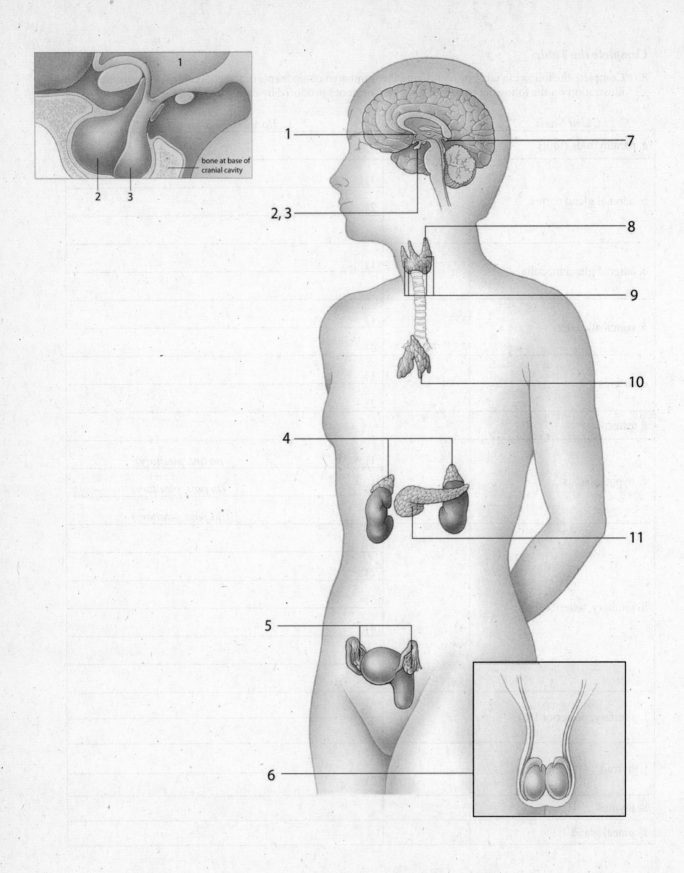

bone at base of
cranial cavity

Short Answer

9. Regardless of their chemical make-up, how do all hormones cause an effect in a target cell? [p.288] _____

_____

_____

_____

_____

Choice

For questions 10–17, choose from the following: [pp.290-291]

a. steroid hormones   b. nonsteroid hormones

10. _____ Lipid-soluble molecules synthesized from cholesterol; made in adrenal glands, ovaries, and testes

11. _____ Includes amines, peptides, proteins, and glycoproteins

12. _____ May diffuse across plasma membrane of target cell into cytoplasm

13. _____ Often enters a cell's nucleus and interacts directly with a gene

14. _____ Hormones that often activate second messengers

15. _____ Peptide hormones that bind to receptors at the plasma membrane

16. _____ Acts by forming a hormone-receptor complex that interacts with specific DNA regions to stimulate or inhibit transcription of mRNA

17. _____ Involves molecules such as cyclic AMP that activate many enzymes in the cytoplasm that, in turn, cause alteration in some cell activity

15.3. THE HYPOTHALAMUS AND PITUITARY GLAND [PP.292-293]

15.4. HORMONES AS LONG-TERM CONROLLERS [P.294]

15.5. GH GROWTH FUNCTIONS AND DISORDERS [P.295]

Boldfaced Terms

hypothalamus _____

_____

pituitary gland _____

_____

gigantism _____

_____

acromegaly _____

_____

pituitary dwarfism _____

_____

Choice-Matching

Label each hormone given in the following list with an "A" if it is secreted by the anterior lobe of the pituitary and a "P" if it is released from the posterior pituitary. Complete the exercise by entering the letter of the corresponding target and hormone action in the parentheses following each label. [pp. 292-293]

1. ____ ( ) ACTH

2. ____ ( ) ADH

3. ____ ( ) FSH

4. ____ ( ) GH (STH)

5. ____ ( ) LH

6. ____ ( ) Oxytocin

7. ____ ( ) Prolactin

8. ____ ( ) TSH

A. Acts on ovaries and testes to produce gametes

B. Acts on mammary glands to stimulate and sustain milk production

C. Acts on ovaries and testes to release gametes; promotes testosterone secretion in males and formation of corpus luteum in females

D. Induces uterine contractions and milk movement into secretory ducts

E. Acts on the thyroid gland to stimulate release of thyroid hormones

F. Acts on the kidneys to induce water conservation and control extracellular fluid volume

G. Acts on the adrenal cortex to stimulate release of adrenal steroid hormones

H. Acts on most cells to promote growth in young; induces protein synthesis and cell division; plays roles in glucose and protein metabolism

Dichotomous Choice

Circle one of two possibilities given between parentheses in each statement. [pp. 292-295]

9. The (hypothalamus / pituitary gland) monitors internal organs and activities related to their functioning, such as eating, sexual behavior, and body temperature; it also secretes some hormones.

10. The (posterior / anterior) lobe of the pituitary stores and secretes two hypothalamic hormones, ADH and oxytocin.

11. The (posterior / anterior) lobe of the pituitary produces and secretes its own hormones, which govern the release of hormones from other endocrine glands.

12. Most hypothalamic hormones acting in the anterior pituitary lobe are (releaser / inhibitor) hormones, and cause target cells to secrete their own hormones.

13. Some hypothalamic hormones slow down hormone secretion from their targets; these are classed as (releaser / inhibitor) hormones.

14. The "cuddle hormone" is (oxytocin / prolactin).

15. The "metabolic hormone" is (TSH / GH).

16. (Pituitary dwarfism / Gigantism) results when not enough somatotropin is produced during childhood.

17. Production of excessive amounts of somatotropin during childhood results in pituitary (dwarfism / gigantism).

236 Chapter Fifteen

18. Excess somatotropin production during adulthood results in thicker bone, cartilage, and connective tissues of hands, feet, jaws, and epithelia; this condition is known as (gigantism / acromegaly).

15.6. THE THYROID AND PARATHYROID GLANDS [PP.296-297]

15.7. ADRENAL GLANDS AND STRESS RESPONSES [PP.298-299]

15.8. THE PANCREAS: REGULATING BLOOD SUGAR [P.300]

15.9. BLOOD SUGAR DISORDERS [P.301]

Boldfaced Terms

thyroid gland _____

_____

simple goiter _____

_____

Grave's disease _____

_____

parathyroid glands_____

_____

rickets _____

_____

adrenal cortex _____

_____

glucocorticoids _____

_____

gluconeogenesis _____

_____

hypoglycemia _____

_____

mineralocorticoids_____

_____

adrenal medulla _____

_____

pancreatic islet _____

_____

diabetes mellitus _____

_____

metabolic acidosis _____

_____

type 1 diabetes _____

_____

type 2 diabetes _____

_____

hypoglycemia _____

_____

Matching

Match the following glands/organs and the hormone(s)/substances produced. Place the letter of the correct gland/organ on the blank line. [pp. 296-301]

Hormone

1. _____ thyroxine and triiodothyronine
2. _____ glucagon
3. _____ PTH
4. _____ somatostatin
5. _____ digestive enzymes
6. _____ glucocorticoids (including cortisol)
7. _____ epinephrine
8. _____ insulin
9. _____ mineralocorticoids (including aldosterone)
10. _____ calcitonin
11. _____ norepinephrine

Gland/Organ

A. adrenal cortex
B. adrenal medulla
C. thyroid
D. parathyroids
E. pancreas (alpha cells)
F. pancreas (beta cells)
G. pancreas (delta cells)
H. pancreas

Short Answer

12. Describe the process by which a simple goiter develops. [p.296] _____

_____

_____

13. What are the symptoms and causes of Grave's disease? [p.296] _____

_____

_____

14. Describe the process of gluconeogenesis. [p.298] _____

_____

_____

15. What are some causes of hypoglycemia? What are some possible results? [p.301] _____

_____

_____

Fill-in-the-Blanks [pp. 296-297]

Thyroxine (T4) and triiodothyronine (T3) are the main hormones secreted by the human (16)

_____ gland. They affect a person's overall (17) _____ rate, growth, and development. The

thyroid also makes (18) _____, a hormone that lowers the level of calcium (and phosphate) in the

blood. The synthesis of thyroid hormones requires (19) _____, which is obtained from the diet. In the

absence of iodine, blood levels of these hormones decrease. The anterior pituitary responds by secreting (20)

_____. Excess TSH over stimulates the thyroid gland and causes it to enlarge. This tissue enlargement

leads to an enlargement of the gland called simple (21) _____. Insufficient thyroid output is called (22)

_____. Hypothyroid adults tend to be (23) _____ [p.296], sluggish, intolerant of cold, and

sometimes feel confused and depressed. Simple goiter is no longer common in areas where people use (24)

_____. When blood levels of thyroid hormones become too high, (25) _____ disease or (26)

_____ goiter may develop. Conditions like these that are caused by elevated levels of thyroid hormones

are called (27) _____. Symptoms include (28) _____ heart rate, profuse sweating and

elevated blood pressure.

The (29) _____ glands are four glands located on the back of the thyroid. They secrete (30)

_____ hormone (PTH), the main regulator of the (31) _____ level in blood. PTH is the

hormone in charge of bone (32) _____. PTH stimulates the reabsorption of calcium from the filtrate

flowing through the (33) _____. In addition, it helps to activate vitamin (34) _____, which

improves the absorption of calcium from food in the GI tract. In vitamin D deficiency, too little calcium and

phosphorus are absorbed, so bones develop improperly. This ailment is called (35) _____. In

The Endocrine System   **239**

hyperparathyroidism, PTH causes too much calcium to be withdrawn from the (36) _____, which

weakens bone tissue. In addition, the excess calcium in the blood may cause (37) _____ and

abnormally functioning (38) _____.

Matching

Choose the most appropriate definition for each term.   [p.298]

39. _____   glucocorticoids

40. _____   cortisol

41. _____   gluconeogenesis

42. _____   hypoglycemia

43. _____   mineralocorticoids

44. _____   aldosterone

45. _____   fight-flight response

46. _____   adrenal medulla

47. _____   adrenal cortex

A. Ongoing low blood glucose level; can develop due to cortisol deficiency

B. Type of adrenal hormones that adjust the concentrations of mineral salts in the extracellular fluid

C. Outer part of each adrenal gland

D. Effects of epinephrine and norepinephrine including increased heart rate and enhanced respiration

E. Type of adrenal hormones that raise the blood level of glucose

F. Synthesis of glucose from amino acids

G. Most abundant mineralocorticoid

H. Body's main glucocorticoid

I. Inner part of each adrenal gland

Short Answer

48. What effect does "bad" stress have on the immune system, and what is a probable result? [p.299] _____

_____

_____

_____

Dichotomous Choice

Circle the correct choice of the two given between parentheses in each statement. [pp. 300-301]

49. The pancreas has the (exocrine / endocrine) function of secreting digestive enzymes, and the (exocrine / endocrine) function of secreting hormones.

50. The endocrine cells of a pancreas are found in clusters called pancreatic (islets / patches).

51. Glucagon is secreted by (alpha / beta) cells.

52. The effect of glucagon is to (raise / lower) the glucose level in the blood.

53. Insulin is secreted by (alpha / beta) cells.

54. The effect of insulin is to (raise / lower) the glucose level in the blood.

55. Somatostatin secreted by delta cells acts on beta and alpha cells to (stimulate / inhibit) the secretion of insulin and glucagon.

56. Insulin deficiency can lead to diabetes mellitus, a disorder in which the blood glucose level (rises / decreases) and glucose accumulates in the urine.

57. In a person with diabetes mellitus, water loss through urination is (reduced / excessive), so the body's water–solute balance becomes disrupted.

58. Lacking a steady glucose supply, body cells of a person with diabetes mellitus begin breaking down fats and proteins for (energy / water).

59. In diabetes mellitus, metabolic acidosis may develop in which blood pH is dangerously (increased / decreased).

60. The trigger for type 1 diabetes may be (obesity / a viral infection).

61. In (type 1 diabetes / type 2 diabetes) the body mounts an autoimmune response against its own insulin-secreting beta cells and destroys them.

62. Juvenile-onset diabetes is also known as (type 1 diabetes / type 2 diabetes).

63. In (type 1 diabetes / type 2 diabetes), insulin levels are close to or above normal, but target cells fail to respond to insulin.

64. A huge risk factor for type 2 diabetes is (obesity / a viral infection).

65. Elevated blood sugar damages (white blood cells / capillaries).

66. Poor circulation results in tissue (edema / death) leading to blindness, amputation, and severe kidney problems.

67. Metabolic syndrome refers to a group of characteristics that indicate a person (has / is at risk for) type 2 diabetes.

68. Once a person is "prediabetic", proper diet and exercise (can / cannot) prevent worsening of the situation.

15.10 OTHER HORMONE SOURCES [P.302]

15.11 THE ENDOCRINE SYSTEM IN HOMEOSTASIS [P.303]

Boldfaced Terms

gonads _____

_____

pineal gland _____

_____

biological clock _____

_____

seasonal affective disorder (SAD) _____

_____

Short Answer

1. What is the function of testosterone in the female and where is it produced? [p.302] _____

_____

_____

2. What is the function of estrogen and progesterone in the male and where are they produced? [p.302]_____

_____

_____

3. What is the function of melatonin and how is its secretion affected by day length? [p.302]_____

_____

4. How do the heart's atria control blood pressure? [p.302]_____

_____

_____

Matching

In questions 5-10, place the letter of the correct definition on the blank line next to the system. [p.303]

5. ____ muscular system

6. ____ digestive system

7. ____ urinary system

8. ____ nervous system

9. ____ reproductive system

10. ____ skeletal system

A. Insulin and GH stimulate cells to take up glucose from the bloodstream

B. Oxytocin works with prolactin to bring about milk release for a nursing infant

C. Parathyroid hormone adjusts blood calcium and potassium levels for muscle contraction

D. Calcitonin stimulates uptake of calcium from the blood to form bone

E. Epinephrine is involved in the fight-flight response

F. Aldosterone and ANP are involved in the management of the body's salt-water balance

SELF-QUIZ

Multiple Choice

____ 1. The anterior lobe of the _____ governs the release of hormones from other endocrine glands, while the posterior lobe stores hormones secreted by the _____. [p.292]

a. pituitary; hypothalamus
b. pancreas; hypothalamus
c. thyroid; parathyroid glands
d. hypothalamus; pituitary
e. pituitary; thalamus

____ 2. ADH is sometimes called vasopressin because it increases _____. [p.293]

a. heart rate
b. sweat production
c. water excretion
d. blood vessel diameter
e. blood pressure

____ 3. The anterior lobe of the pituitary secretes _____ different hormones, while the posterior lobe releases _____ hormones. [pp.292-293]

a. two, six
b. six, six
c. six, none
d. six, two
e. two, two

___ 4. If calcium were eliminated from your diet, your body would secrete more _____ in an effort to release calcium stored in your bones and send it to the tissues that require it. [p.297]

    a. parathyroid hormone
    b. aldosterone
    c. calcitonin
    d. mineralocorticoids
    e. none of the above

Choice

For questions 5–7, choose from the following answers:

    a. PTH
    b. cortisol
    c. aldosterone
    d. calcitonin
    e. melatonin

___ 5. lowers the level of calcium and phosphate in the blood [p.296]

___ 6. stimulates kidney to reabsorb sodium ions and excrete potassium ions [p.298]

___ 7. affects sleep/wake cycles [p.302]

For questions 8–10, choose from the following answers:

    a. adrenal medulla
    b. adrenal cortex
    c. thyroid
    d. anterior pituitary
    e. posterior pituitary

___ 8. The _____ produces glucocorticoids that help maintain the blood level of glucose and suppress inflammatory responses. [p.298]

___ 9. The gland that is directly associated with the fight-flight response is the _____. [p.298]

___ 10. The _____ gland regulates the basic metabolic rate. [p.296]

Matching

Choose the most appropriate description for each term.

11. ___ ACTH [p.293]

12. ___ ADH [pp.292-293]

13. ___ calcitonin [p.296]

14. ___ cortisol [p.298]

15. ___ epinephrine and norepinephrine [pp.298-299]

16. ___ estrogen [p.302]

17. ___ GH (STH) [p.293]

18. ___ glucagon [p.300]

19. ___ insulin [p.300]

20. ___ melatonin [p.302]

21. ___ oxytocin [p.293]

22. ___ parathyroid hormone [pp.296-297]

23. ___ progesterone [p.302]

24. ___ testosterone [p.302]

25. ___ thymosins [p.302]

The Endocrine System   **243**

26. _____ thyroxine [p.296]

27. _____ TSH [p.293]

A. Raises the glucose level in the blood

B. Influences daily biorhythms, gonad development, and reproductive cycles

C. Affects development of male sexual traits, required for sperm formation

D. Increases heart rate and controls blood volume; the "emergency hormones"

E. Essential for egg maturation and maintenance of secondary sexual characteristics in the female

F. The water-conservation hormone; released from posterior pituitary

G. Lowers blood sugar by signaling cells to take in glucose; promotes synthesis of proteins and fats

H. Stimulates adrenal cortex to secrete steroid hormones

I. Causes increase of calcium in the blood

J. Influences overall metabolic rate, growth, and development

K. Roles in immunity

L. Triggers uterine contractions during labor and causes milk release during nursing

M. Prepares and maintains uterine lining for pregnancy; stimulates breast development

N. Promotes conversion of protein to glucose; primary glucocorticoid

O. Lowers calcium levels in blood

P. Secreted by anterior pituitary; stimulates release of thyroid hormones

Q. Secreted by anterior pituitary; enhances growth in young animals, especially cartilage and bone

CHAPTER OBJECTIVES / REVIEW QUESTIONS

1. Where is a hormone made, how does it travel to its target, and what happens when it arrives there? [p.288]

2. Collectively, sources of hormones are referred to as the _____ system. [p.288]

3. Name the four categories of hormones and give two representative examples of each category . [p.290]

4. Which hormones are released by the posterior pituitary gland and what are their functions. [p.292-293]

5. Contrast the proposed mechanisms of hormonal action on target cell activities by (a) steroid hormones and (b) nonsteroid hormones. [pp.290-291]

6. State the relationship between both the anterior and posterior pituitary lobes and the hypothalamus. [pp.292-293]

7. Which conditions are associated with an imbalance in growth hormone (GH) and what are their characteristics? [p.295]

8. Which conditions are associated with an imbalance in thyroid hormones and what are their characteristics? [p.296]

9. Most hypothalamic hormones acting in the anterior lobe are _____ that cause target cells to secrete hormones of their own. Some are _____ that slow down secretion from their targets. [p.293]

10. Which hormones are produced by the anterior pituitary gland and what are their functions. [pp.292-293]

11. Distinguish between type 1 and type 2 diabetes, including triggering factor and insulin production. [p.301]

12. The most abundantly produced mineralocorticoid is _____; cite its function. [p.298]

13. Describe the ailment called *rickets* and cite its cause. [p.297]

14. The pineal gland secretes the hormone _____; give two examples of the action of this hormone. [p.302]

INTEGRATING AND APPLYING KEY CONCEPTS

1) Suppose you suddenly quadruple your already high daily consumption of calcium. State which organs would be affected and tell how they would be affected. Name two hormones whose levels would most probably be affected, and tell whether your body's production of them would increase or decrease. Suppose you continue this high rate of calcium consumption for ten years. Can you predict the organs that would be subject to the most stress as a result?

2) Imagine that you are a neurologist and a patient is referred to your office with a diagnosed pituitary gland tumor in the anterior lobe. What hormones might be affected and what signs and symptoms would you expect to see in your patient?

16

REPRODUCTIVE SYSTEMS

INTRODUCTION

The consequences of not understanding sexually transmitted disease and birth control can be long-lasting. Studying this chapter will provide an understanding of these issues by first examining the anatomy of male and female reproductive systems, as well as the hormone driven processes of sperm and egg production. You will then apply this information to the topics of birth control, infertility options and STDs.

FOCAL POINTS

- Figures 16.1 [p.308-309] and 16.4 animated [pp.312] provide a starting point for the study of human reproduction. Figure 16.1 is an illustration of the parts of the female reproductive system. Figure 16.4 is the same for a male.
- Figure 16.3 animated [p.311] illustrates the menstrual cycle as controlled by hormones in the female. Understanding this cycle is essential to the understanding of reproduction. This continues the theme that you have seen before of interactions among body systems. You should also notice that the same hormones have different effects on male vs. female reproduction.
- Figure 16.8 [p.315] presents the relationship between the endocrine and reproductive system in the male. This, once again, demonstrates the interactions between body systems.
- Figure 16.10 animated [p.317] illustrates the process of fertilization.
- Table 16.3 [p.318] lists the common methods of contraception and their failure rates.

INTERACTIVE EXERCISES

CHAPTER INTRODUCTION [p.307]

16.1. THE FEMALE REPRODUCTIVE SYSTEM [pp.308-309]

16.2. THE OVARIAN CYCLE: OOCYTES DEVELOP [pp.310-311]

Boldfaced Terms

ovaries _____

_____

germ cells _____

_____

gametes _____

_____

secondary sexual traits _____

_____

estrogens _____

_____

oocytes _____

_____

oviduct _____

_____

uterus _____

_____

endometrium _____

_____

cervix _____

_____

vagina _____

_____

menstrual cycle _____

_____

menstruation _____

_____

progesterone _____

_____

menarche _____

_____

menopause _____

_____

endometriosis _____

_____

ovarian cycle _____

_____

follicle _____

_____

zona pellucida _____

_____

secondary oocyte _____

_____

ovulation _____

_____

corpus luteum _____

_____

Fill-in-the-Blanks

The numbered items on the accompanying illustration represent missing information; complete the numbered blanks in the following narrative to supply the missing information on the illustration. [p.308]

An oocyte (immature egg) is released from a(n) (1) _____ . When the oocyte is released from either ovary, it moves into a(n) (2) _____ and is transported to the (3) _____ , a hollow, pear-shaped organ in which a baby can grow and develop. The wall of the uterus consists of a thick layer of smooth muscle, the (4) _____ , and an interior lining, the (5) _____ , which includes epithelial tissue, connective tissue, glands, and blood vessels. The lower portion of the uterus is the (6) _____ . A muscular tube, the (7) _____ , extends from the cervix to the body surface and receives the penis and sperm and functions as part of the (8) _____ . Outermost is a pair of fat-padded skin folds, the (9) _____ . Those folds enclose a smaller pair of skin folds, the (10) _____ that are highly vascularized but have no fatty tissue. The smaller folds partly enclose the (11) _____ , a small organ sensitive to stimulation . The opening of the (12) _____ is about midway between the clitoris and the vaginal opening.

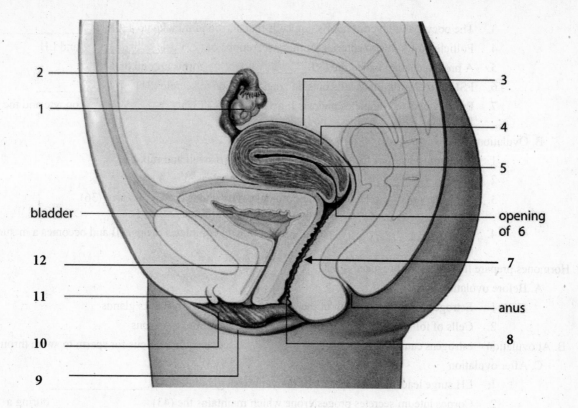

2 _____

1 _____

bladder _____

12 _____

11 _____

10 _____

9 _____

3 _____

4 _____

5 _____

opening of 6 _____

7 _____

anus _____

8 _____

Outlining

Fill in the outline below covering the reproductive cycle of female humans.

I. The menstrual cycle advances through (13) _____ phases. [pp.308-309]

    A. Menstrual phase

        1. (14) _____ - marks the first day of a new cycle

        2. (15) _____ disintegration

        3. oocyte maturation

    B. (16) _____ phase

        1. (17) _____ begins to thicken again

        2. (18) _____ – release of an oocyte from an ovary

    C. (19) _____ phase

        1. (20) _____ forms

        2. (21) _____ is primed for pregnancy by progesterone and estrogen

    D. Terms related to the menstrual cycle

        1. (22) _____ - first menstruation, between ages ten and sixteen

        2. (23) _____ - menstrual cycles stop, occurs in late 40s or early 50s

        3. (24) _____ - a disorder in which endometrial tissue spreads outside of uterus

II. The (25) _____ cycle includes maturation of a primary oocyte and ovulation. [pp.310-311]

    A. Primary oocyte maturation

        1. Primary oocytes enter (26) _____ which is halted before the first division is complete.

        2. Primary oocytes are found near the surface of an (27) _____

Reproductive Systems   **249**

3. The oocyte and layer of nourishing cells that surround it make up a (28) _____

4. Follicle grows due to anterior pituitary secretions of (29) _____ and LH

5. A protein coating called the (30) _____ forms around oocyte

6. FSH and LH stimulate cells outside zona pellucida to secrete (31) _____

7. Primary oocyte completes meiosis I, giving rise to a (32) _____ oocyte and the first (33) _____ body

  B. Ovulation

1. Surge of LH causes the (34) _____ to swell and rupture

2. (35) _____ and first polar body are released

3. Secondary oocyte released into abdominal cavity and is drawn into a(n) (36) _____ by the beating of the cilia or fimbriae

4. If (37) _____ takes place, the oocyte completes meiosis II and becomes a mature (38) _____

III. Hormones prepare the uterus for pregnancy [p.311]

  A. Before ovulation

1. Estrogens stimulate growth of the (39) _____ and its glands

2. Cells of follicle wall secrete (40) _____ and estrogens

  B. At ovulation – estrogens cause the (41) _____ to secrete a thin mucus for sperm to swim through

  C. After ovulation

1. LH surge leads to development of the yellowish glandular (42) _____

2. Corpus luteum secretes progesterone which maintains the (43) _____ during a pregnancy

3. While the corpus luteum persists, the hypothalamus signals a decrease in FSH preventing other (44) _____ from developing

4. Corpus luteum breaks down after about twelve days if an embryo doesn't (45) _____ in the endometrium

5. When progesterone and estrogen levels drop, the (46) _____ breaks down and menstruation occurs

6. The cycle begins again as rising levels of (47) _____ stimulate the repair and growth of the endometrium

Analyzing Diagrams

Study the accompanying diagram to correctly complete the following dichotomous choice statements. [p.311]

48. With increasing levels of FSH and LH secreted by the pituitary, a follicle (shrinks / grows).

49. Ovulation occurs when there is a sharp surge of (FSH / LH) levels.

50. FSH and LH levels (decrease / increase) after ovulation.

51. As a follicle develops (prior to ovulation), estrogen levels (decrease / increase).

52. After ovulation, estrogen levels drop slightly, while progesterone levels (decrease / increase).

53. As long as the corpus luteum remains, estrogen and progesterone levels (remain stable / continue to increase).

54. As long as the corpus luteum remains, the endometrium is (very thin / fully developed).

55. The endometrium is fully developed during the (follicular / luteal) stage of the menstrual cycle.

56. The menstrual cycle begins with (ovulation / menstruation).

57. Progesterone appears to cause the endometrium to (deteriorate / be maintained).

58. Estrogen production continues throughout the existence of a follicle, while progesterone is only made in large amounts after the formation of the (corpus luteum / endometrium).

59. Fertilization is possible (throughout / around the middle of) the menstrual cycle.

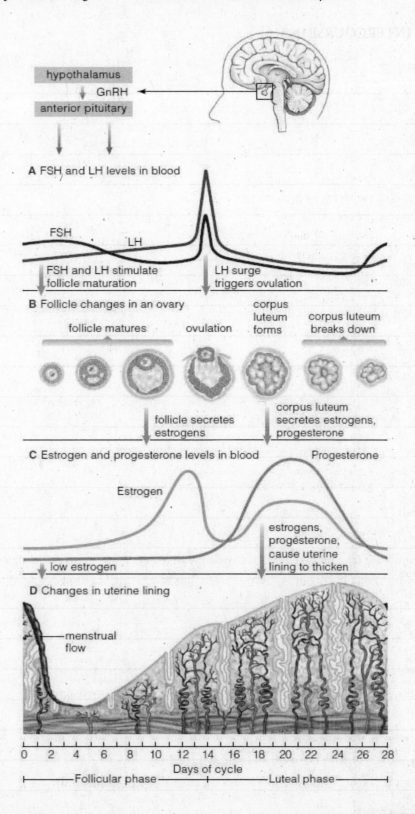

16.3. THE MALE REPRODUCTIVE SYSTEM [pp.312-313]

16.4. HOW SPERM FORM [pp.314-315]

16.5. SEXUAL INTERCOURSE [p.316]

Boldfaced Terms

testes _____

_____

epididymis _____

_____

vas deferens _____

_____

penis _____

_____

semen _____

_____

seminal vesicles _____

_____

prostate gland _____

_____

bulbourethral glands _____

_____

seminiferous tubules _____

_____

spermatogenesis_____

_____

sperm _____

_____

Sertoli cells _____

_____

acrosome _____

_____

Leydig cells _____

_____

testosterone _____

_____

coitus _____

_____

orgasm _____

_____

True-False

If the statement is true, write a "T" in the blank. If the statement is false, make it correct by changing the underlined word(s) and writing the correct word(s) in the answer blank. [pp.312-313]

1. _____ Testes are the male <u>gonads</u>.

2. _____ In a male, the testes descend into the scrotum <u>after</u> birth.

3. _____ For sperm to develop properly, the temperature inside the scrotum must be a few degrees <u>warmer</u> than inside the body.

4. _____ When human sperm leave the testes, they are <u>not</u> mature.

5. _____ Until sperm leave the body, they are stored in the <u>scrotum</u>.

6. _____ Semen consists of <u>sperm only</u>.

7. _____ Seminal vesicles secrete <u>fructose</u> and prostaglandins.

8. _____ Secretions from the <u>bulbourethral</u> gland buffer the acidic environment of the urethra caused by traces of urine.

9. _____ There is a <u>pair of</u> prostate gland(s).

Sequence and Labeling

First, write the letter of each structure listed in the order that the sperm travel through them. Put the first location of sperm by number 10, the second by number 11, and so on. As you do this, label the diagram below. [p.312]

    a. epididymis   b. urethra   c. ejaculatory duct   d. vas deferens   e. testis

10. ____

11. ____

12. ____

13. ____

14. ____

Now, place the letter of each gland that contributes to the semen in the order that their products enter the urethra as the sperm passes them prior to ejaculation. As you do, label the diagram to the right.

a. bulbourethral gland   b. seminal vesicle   c. prostate gland

15. ____

16. ____

17. ____

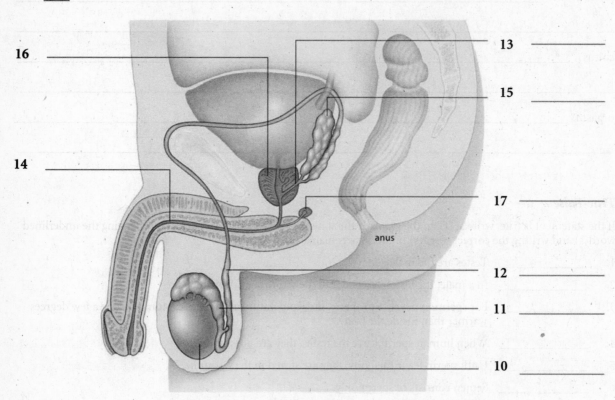

16 _____

14 _____

13 _____

15 _____

17 _____

12 _____

11 _____

10 _____

anus

Fill-in-the-Blanks . [p.314]

Each testis is packed with over 400 feet of (18) _____. Inside their walls are cells called (19)

_____ . During spermatogenesis, these cells undergo divisions, including a type called (20)

_____ and a type called (21) _____ . This process results in the specialized male

reproductive cells called *sperm*. Spermatogonia develop into (22) _____ , which after a meiotic division

known as (23) _____ are termed (24) _____ . A second division known as (25)

_____ results in immature sperm known as (26) _____ which gradually develop into

spermatozoa, or simply (27) _____ , the male gametes. The "tail" or (28) _____ of each sperm arises at the very end of the process, which takes 9 to 10 weeks. These developing sperm cells receive

nourishment and chemical signals from adjacent (29) _____ cells. A mature sperms has a tail, a

midpiece, and a(an) (30) _____. Within the head is a nucleus containing DNA organized into

chromosomes. An enzyme-containing cap, the (31) _____, covers most of the head. Its enzymes

help sperm penetrate the extracellular material around an egg at fertilization. In the midpiece

(32) _____ supply energy for the tail's whip-like movements.

Dichotomous Choice

Circle one of two possibilities given between parentheses in each statement. [p.315]

33. Interstitial cells are also known as (Leydig cells/ Sertoli cells).

34. Testosterone is secreted by (Leydig cells / the hypothalamus).

35. (Testosterone / FSH) governs the growth, form, and functions of the male reproductive tract.

36. Sexual behavior and secondary sexual traits are associated with (LH / testosterone).

37. LH and FSH are secreted by the (anterior / posterior) lobe of the pituitary gland.

38. The (testes / hypothalamus) govern(s) sperm formation by controlling interactions among testosterone, LH, and FSH.

39. When blood levels of testosterone (increase / decrease) beyond a certain set point, the hypothalamus secretes GnRH, which stimulates the anterior pituitary lobe to release LH and FSH, which the bloodstream distributes to the testes.

40. Within the testes, (LH / FSH) acts on Leydig cells stimulating them to secrete testosterone.

41. Sertoli cells have receptors for (LH / FSH) which is crucial to commencing spermatogenesis at the time of puberty.

42. When blood testosterone levels (increase / decrease) past a set point, feedback loops to the hypothalamus slow down testosterone secretion and sperm formation.

Elimination

From the choices within parentheses, select the word that does not correctly complete the statement. [p.316]

43. Sexual intercourse is technically called (coitus / fertilization / copulation).

44. The male sex act involves (meiosis / erection / ejaculation).

45. Vasodilation occurs in (the male's penis / the female's genital area / the male's bladder).

46. Semen consists of sperm combined with secretions of the (pancreas / prostate gland / seminal vesicles).

47. The culmination of the sex act usually involves (ejaculation / orgasm / pregnancy).

48. Pregnancy in a female requires (orgasm / sperm / secondary oocyte).

49. Fertilization may result from sperm entering the vagina (a few days before ovulation / a few days after ovulation / at any time in the menstrual cycle).

16.6. FERTILIZATION [p.317]

16.7. PREVENTING PREGNANCY [pp.318-319]

16.8. OPTIONS FOR COPING WITH INFERTILITY [pp.320-321]

Boldfaced Terms

fertilization _____

_____

zygote _____

_____

ovum _____

_____

abortion _____

_____

in vitro fertilization (IVF) _____

_____

Matching

Match each birth control option with its description . [pp.318-319]

1. _____ abstinence
2. _____ rhythm or sympto-thermal method
3. _____ withdrawal
4. _____ douching
5. _____ vasectomy
6. _____ tubal ligation
7. _____ spermicides
8. _____ diaphragm
9. _____ cervical cap
10. _____ contraceptive sponge
11. _____ IUD
12. _____ condoms
13. _____ oral contraceptive (the "pill")
14. _____ morning-after pills
15. _____ abortiom
16. _____ birth control patch

A. Each vas deferens is severed and tied off

B. Removal of the penis from the vagina before ejaculation

C. A small plastic or metal device that is placed in the uterus and interferes with implantation

D. Rinsing out the vagina with a chemical after intercourse

E. The removal of an embryo or fetus from the womb.

F. No sexual intercourse

G. The oviducts are cauterized or cut and tied off

H. Hormone delivered through the skin; blocks ovulation

I. A soft, disposable disk that contains a spermicide and covers the cervix; inserted up to 24 hours before intercourse

J. Widely used method of fertility control; contains synthetic estrogens and progesterones; suppresses maturation and release of oocytes

K. Avoiding intercourse during the woman's fertile period; uses daily temperature readings

L. Thin, tight-fitting sheaths of latex or animal skin worn over the penis during intercourse

M. Flexible, dome-shaped device that is inserted into the vagina and positioned over the cervix before intercourse

N. Toxic to sperm; packaged in an applicator and placed in the vagina just before intercourse; not reliable unless used with a diaphragm or condom

O. Variation on the diaphragm; smaller and can be left up to 3 days with a single dose of spermicide

P. Emergency contraception to suppress ovulation; must be taken within 5 days of intercourse

Short Answer

17. After what stage of pregnancy are abortions particularly controversial? [p.319] _____

_____

_____

_____

Fill-in-the-Blanks [p.317]

After intercourse, contractions of smooth muscle in the (18) _____ help move the sperm toward the oviducts. (19) _____, a weakening of the membrane covering the acrosome, begins. When an oocyte is encountered, enzymes released from the acrosome eat through the (20) _____ surrounding the oocyte. Penetration of the sperm triggers the oocyte to complete (21) _____ . This results in a mature (22) _____. (23) _____ is the fusion of the nuclei of an egg and a sperm. The resulting cell is called a (24) _____.

Fill-in-the-Blanks [p.320-321]

(25) _____ of infertility cases are due to problems with the oocyte or ovulation. These women may be treated with (26) _____ drugs or (27) _____ to stimulate ovulation. (28) _____, using concentrated sperm from a partner or donor, was one of the first methods of assisted reproductive technology. In (29) _____, eggs and sperm are introduced in a laboratory dish. A few days later, the young embryos are implanted in a woman's (30) _____. In (31) _____, a sperm is mechanically injected into an egg in a laboratory. In (32) _____ or (33) _____, gametes or lab-created zygotes are introduced into a woman's oviducts. In vitro fertilization, GIFT and ZIFT produce a normal pregnancy about (34) _____ % of the time.

16.9. SOME COMMON SEXUALLY TRANSMITTED DISEASES [pp.322-323]

16.10. STDS CAUSED BY VIRUSES AND PARASITES [pp.324-325]

16.11. EIGHT STEPS TO SAFER SEX [p.325]

16.12. CANCERS OF THE BREAST AND REPRODUCTIVE SYSTEM [pp.326-327]

Boldfaced Terms

sexually transmitted diseases (STDs) _____

_____

chlamydia _____

_____

pelvic inflammatory disease (PID) _____

_____

gonorrhea _____

_____

syphilis _____

_____

genital herpes _____

_____

genital warts _____

_____

hepatitis B _____

_____

hepatitis C _____

_____

pubic lice _____

_____

candidiasis _____

_____

trichomoniasis _____

_____

breast cancer _____

_____

Reproductive Systems **259**

endometrial cancer _____

_____

ovarian cancer _____

_____

testicular cancer _____

_____

prostate cancer _____

_____

Choice

For questions 1–12, choose the letter that applies from the following: [pp.322-323]

   a. gonorrhea   b. syphilis   c. chlamydial infection

1. ____   At least 30% of newborns with eye infections and pneumonia were infected with this during birth.

2. ____   Can infect epithelial cells of genital tract, rectum, eye membranes and throat

3. ____   Leads to PID in 20 to 40% of infected women

4. ____   First sign is an ulcer called a chancre

5. ____   Reinfection is possible due to presence of many different strains.

6. ____   Secondary stage involves lesions in mucous membranes, eyes, bones and CNS, along with a rash over most of the body.

7. ____   Bacteria enter lymph nodes causing swelling in nodes and surrounding area.

8. ____   Tertiary stage begins 5 to 20 years after infection and involves lesions and scarring in many parts of the body.

9. ____   Responsible for miscarriages and stillbirths.

10. ____   Treponemes will eventually damage brain and spinal cord, leading to insanity and paralysis.

11. ____   Symptoms include discharge from penis or vagina, and burning sensation upon urination.

12. ____   Symptoms in males include a discharge of pus from penis; sterility possible if untreated.

Short Answer

13. What does PID stand for? What STDs may lead to PID? What are its symptoms? [p.322] _____

_____

_____

_____

_____

Matching

Choose the most appropriate description for each term. [pp.324-325]

14. ____ hepatitis B
15. ____ herpes simplex virus
16. ____ human papilloma virus (HPV)
17. ____ trichomoniasis
18. ____ pubic lice
19. ____ hepatitis C
20. ____ candidiasis

A. virus causing severe liver cirrhosis and even cancer, can be transmitted through contaminated blood

B. present in 20% of population over 12; outbreaks produce recurring genital sores

C. foul-smelling discharge from vagina or penis with itching and burning; caused by a protozoan

D. presence of these blood-feeding animals in pubic hair causes intense irritation

E. highly contagious liver virus transmitted through saliva, vaginal secretions, semen or blood

F. vaginal infection caused by yeast

G. causes genital warts; some forms cause most cases of invasive cervical cancer

Fill-in-the-Blanks [pp.326-327]

In women, (21) _____ cancer is the second most common cause of death due to cancer. Risk is increased by the presence of a faulty (22) _____ or _____ gene. The American Cancer Society recommends an annual (23) _____ for women over 40 in order to detect small cancers which can often be removed by a simple (24) _____. A (25) _____ removes the breast, chest muscles and associated lymph nodes. Cancers of the uterus most often affect the (26) _____ and (27) _____. (28) _____ cancer is often lethal because symptoms do not appear until the cancer is advanced. (29) _____ cancer in men can be caught early by monthly self-exams. The second most deadly cancer in men is (30) _____ cancer.

SELF-QUIZ

Choice

For questions 1–3, choose from the following answers:

a. *Leydig (or interstitial) cells*   b. *seminiferous tubules*   c. *Sertoli cells*   d. *epididymis*

_____ 1. Sperm mature and are stored in the _____. [p.313]

_____ 2. Testosterone is produced by the _____. [p.315]

_____ 3. Male gametes are produced by meiosis in the _____. [p.314]

For questions 4–5, choose from the following answers:

a. *cervix*   b. *oviduct*   c. *urethra*

_____ 4. The _____ is the lower narrowed portion of the uterus. [p.308]

_____ 5. The _____ is a pathway from the ovary to the uterus. [p.308]

For questions 6–9, choose from the following answers:

a. *corpus luteum*   b. *follicle* c. *pituitary*   d. *hypothalamus*

_____ 6. Provides a midcycle surge of LH to trigger ovulation [p.311]

_____ 7. Secretes follicle-stimulating hormone (FSH) and luteinizing hormone (LH) [p.310]

_____ 8. Secretes GnRH, which stimulates the pituitary to begin secreting LH and FSH [p.310]

_____ 9. Secretes some estrogen and progesterone [p.311]

For questions 10–12, choose from the following answers:

a. *Sympto-thermal or rhythm*   b. *Diaphragm*   c. *IUD*   d. *Spermicides*

_____ 10. _____ is a flexible, dome-shaped device that is inserted into the vagina just before intercourse. [p.318]

_____ 11. _____ are toxic to sperm, packaged in an applicator, and placed in the vagina just before intercourse. [p.318]

_____ 12. _____ is the avoidance of intercourse during the woman's fertile period. [p.318]

For questions 13-17, choose the letter that applies from the following:

a. *gonorrhea*   b. *herpes*   c. *chlamydial infection*   d. *hepatitis B*

e. *pelvic inflammatory disease*   f. *human papilloma virus*

_____ 13. This infection damages the liver and is far more contagious than HIV. [p.324]

_____ 14. Serious complication of some STDs; severe abdominal pain, scarred oviducts leading to abnormal pregnancies and sterility. [p.322]

_____ 15. A bacterium, Neisseria gonorrhea, enters mucous membranes and causes more noticeable symptoms in males; prompt treatment cures this disease. [pp.322-323]

_____ 16. Between flare-ups, the virus remains latent within the nervous system. [p.324]

_____ 17. Causes painless clustered growths on the penis, cervix, or around the anus; probable cause of cervical cancer. [p.324]

CHAPTER OBJECTIVES / REVIEW QUESTIONS

1. The primary reproductive organs are sperm-producing _____ in males and egg-producing _____ in females. [pp.308, 312]

2. Describe the relationships of the uterus, cervix and endometrium. [p.308]

3. Ovulation is the release of a primary _____ from the ovary. [p.311]

4. Name the event that brings about ovulation, and name the other hormonal events that bring about the onset and finish of menstruation. [pp.310-311]

5. Describe the origin and functions of the corpus luteum. [pp.311]

6. The four hormones that control egg maturation and release as well as changes in the endometrium are _____, _____, _____, and _____; they are part of feedback loops involving the hypothalamus, anterior pituitary, and ovaries. [pp.311]

7. In males the primary reproductive organs (testes) also release _____ which influence reproductive functions and _____ sexual traits. [p.315]

8. Where are sperm made? Where do they mature? [pp.314-315]

9. Follow the path of a mature sperm from the seminiferous tubules to the urethral exit. List every structure encountered along the path and state its contribution to the nurture of the sperm [p.313]

10. List, in order, the stages of spermatogenesis. [p.314]

11 Name the three hormones that directly control sperm formation and that form part of feedback loops among the hypothalamus, anterior pituitary, and testes. [p.315]

12. List the physiological events that bring about erection of the penis during sexual stimulation, and explain the process of ejaculation. [p.316]

13. Where does fertilization usually occur? About how many sperm make it to this location? [p.316-317]

14. Two different types of surgical birth control are _____ and _____. [p.318]

15. The _____ is a method of birth control that also helps prevent the spread of sexually transmitted diseases. [p.318]

16. During what portion of a pregnancy is abortion legal in the U.S.? [p.319]

17. Generally define *artificial insemination* and *AID*. [p.320]

18. Describe *in vitro* fertilization as a method of overcoming infertility. [p.320]

19. Distinguish among ZIFT and GIFT as methods of intrafallopian transfers. [p.321]

20. List the names and symptoms of three bacterial STDs that can be sexually transmitted by humans to their partners. [pp.322-323]

21. List the names and symptoms of three viral STDs that can be sexually transmitted by humans to their partners. [p.324]

22. For each of three STDs, state how you can avoid being infected, or, if in spite of all your precautions, you did become infected, what would be the likely course of treatment. [p.322-325]

23. Name and briefly describe STDs transmitted by an animal, a fungus and a protozoan. [p.325]

24. Which two genes are associated with an increased risk of developing breast cancer? [p.326]

INTEGRATING AND APPLYING KEY CONCEPTS

What percentage of humans on the planet today do you think resulted from unplanned pregnancies? As technology allows increasingly better control of fertility, what effect do you think this might have on size of families or age of parents having families? What effect will it have on societies that currently do not have easily available birth control? Assume that technology will continue to be available to some groups or populations before others, what effect do you think this will have on increases or decreases in the relative sizes of populations?

17

DEVELOPMENT AND AGING

INTRODUCTION

This chapter examines the genetically controlled changes that occur from the fertilization of an egg by a sperm, through development beyond birth, and the process of senescence as we age. As you read through the chapter, be aware of the development and maturation of the body systems that you have studied.

FOCAL POINTS

- Figure 17.2 [p.333] diagrams the stages of embryonic development.
- Figure 17.3 animated [p.334] follows the events after fertilization of an egg through implantation in the uterine wall. As new terms are introduced in this section of the chapter, relating them to this illustration will be very helpful.
- Figure 17.15 [p.343] diagrams blood flow in the fetus prior to, and following, birth.
- Figure 17.19 animated [p.346] presents a great deal of valuable information about embryonic and fetal development in a concise table. Using this "big-picture" figure will help you to understand the stages of development in the uterus.
- Table 17.2 [p.349] gives some insight into several changes that take place as we age and may help in the understanding of how humans are physiologically different in our later years than we were in our prime.

INTERACTIVE EXERCISES

CHAPTER INTRODUCTION [p.331]

17.1. OVERVIEW OF EARLY HUMAN DEVELOPMENT [pp.332-333]

17.2. FROM ZYGOTE TO IMPLANTATION [pp.334-335]

17.3 A BABY TIMES TWO [p.335]

Boldfaced Terms

cleavage _____

_____

blastomere _____

_____

gastrulation _____

_____

germ layers _____

_____

cell differentiation _____

_____

morphogenesis _____

_____

blastocyst _____

_____

inner cell mass _____

_____

embryo _____

_____

implantation _____

_____

Sequence

Arrange the following events in correct chronological sequence. Write the letter of the first step next to 1, the letter of the second step next to 2, and so on. [p.332]

1. ____

2. ____

3. ____

4. ____

5. ____

6. ____

A. gastrulation

B. fertilization

C. cleavage

D. cell differentiation

E. morphogenesis

F. gamete formation

Complete the Table

7. Complete the following table by entering the correct embryonic germ layer (ectoderm, mesoderm, or endoderm) that forms the tissues and organs listed. [p.332]

Germ Layer	Tissues/Organs
a.	Muscle
b.	Nervous system
c.	Linings of the digestive tract and lungs
d.	Cardiovascular system (blood vessels, heart)
e.	Epidermis (outer layer of skin)
f.	Reproductive and urinary systems
g.	Pituitary gland
h.	Cartilage and bone
i.	Sense organs

Fill-in-the-Blanks [pp.332-333]

When the nucleus of a sperm and the nucleus of an egg fuse, (8) _____ takes place. This process forms a (9) _____ . As this travels down the oviduct towards the uterus, (10) _____ takes place converting it to a ball of cells. By the time it reaches the uterus it is a cluster of sixteen cells called a(an) (11) _____ . Next comes (12) _____ , a process that rearranges the cells. Once this happens, (13) _____ occurs thus allowing the cells to become specialized. Once the cells are specialized, (14) _____ occurs, a process by which tissues and organs form. This process requires sheets of tissue to (15) _____ and certain cells to (16) _____ on cue.

Matching

Choose the most appropriate description for each term. [pp.332-335]

17. ____ cleavage
18. ____ blastomere
19. ____ gastrulation
20. ____ identical twins
21. ____ fraternal twins
22. ____ morula
23. ____ blastocyst
24. ____ embryo
25. ____ HCG
26. ____ implantation
27. ____ ectopic pregnancy

A. A solid ball of 16 to 32 cells three or four days after fertilization

B. Produced by separation of the early embryo into two; two independent embryos develop

C. About one week after fertilization, epithelial cells of the blastocyst become embedded into the endometrium

D. Develops from the inner cell mass

E. Caused by implantation of a fertilized egg in an oviduct or in the abdominal wall

F. Produced when two different eggs are fertilized at roughly the same time by two different sperm

G. Process of cell division that converts a zygote to a ball of cells

H. Each new cell that forms during cleavage

I. Stimulates the corpus luteum to continue secreting estrogen and progesterone

J. A process that rearranges the cells of the morula

K. A hollow ball of cells that consists of a surface epithelium and an inner clump of cells to one side of the ball

17.4. HOW THE EARLY EMBRYO DEVELOPS [PP.336-337]

17.5. EXTRAEMBRYONIC MEMBRANES [p.338]

17.6 THE PLACENTA: A PIPELINE FOR OXYGEN, NUTRIENTS, AND OTHER SUBSTANCES [p.339]

Boldfaced Terms

embryonic disk _____

_____

neural tube _____

_____

extraembryonic membranes _____

_____

yolk sac _____

_____

amnion _____

_____

allantois _____

_____

umbilical cord _____

_____

chorion _____

_____

placenta _____

_____

Matching

Choose the most appropriate answer for each term. [pp.336-337]

1. ____ embryonic period
2. ____ apoptosis
3. ____ notochord
4. ____ neural tube
5. ____ somites
6. ____ coelom
7. ____ embryonic disc
8. ____ first trimester
9. ____ primitive streak

A. Paired blocks of mesoderm that give rise to most bones and skeletal muscles of neck and trunk plus their dermal coverings
B. 3 month period during most of which a developing baby is called an embryo
C. Appears around day 15 along midline of embryonic disk; first indication of gastrulation
D. The vertebral column will develop around this flexible rod of cells
E. Begins shortly after fertilization and lasts for eight weeks
F. After implantation, the inner cell mass transforms to this structure
G. Genetically programmed destruction of cells to help sculpt body parts
H. Gives rise to the brain and spinal cord
I. Formed by spaces that open up in the mesoderm and then coalesce to form a larger cavity

Labeling

Identify each indicated part of the following illustrations. [p.336]

10. _____   15. _____

11. _____   16. _____

12. _____   17. _____

13. _____   18. _____

14. _____

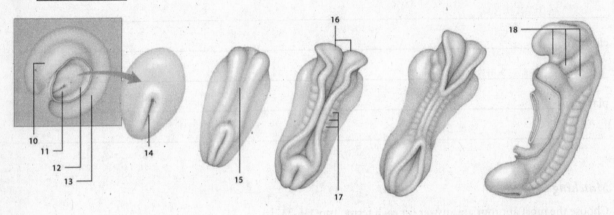

Choice

In the blank beside each structure, write the letter of the embryonic feature that gives rise to it. [pp.336-337]

a. neural tube   b. coelom   c. pharyngeal gill arches   d. notochord

19. ____ face, mouth and neck

20. ____ cavity between the body wall and digestive tract

21. ____ brain and spinal cord

22. ____ vertebral column

Dichotomous Choice

From the pair in parentheses, select the term that best completes the sentence. [p.337]

23. Cells of the (endoderm / ectoderm) at the embryo's midline elongate and form a neural plate.

24. The neural plate folds over and meets at the midline to form the neural (tube / vertebrae).

25. Migrating Schwann cells stick to adhesion proteins on the surface of axons (but not on/ and also on) blood vessels.

26. Migrating cells follow adhesive cues that tell the cells how to find their way and when to (stop / die).

27. Apoptosis is genetically programmed cell (division / death).

Choice

For questions 28–40, choose from the following: [pp.338-339]

   a. yolk sac   b. amnion   c. allantois   d. umbilical cord   e. chorion   f. placenta

28. _____ A close association of the embryonic chorion and the superficial cells of the mother's endometrial lining

29. _____ A protective membrane that surrounds the embryo and other membranes

30. _____ Membrane that continues the secretion of HCG that began when the blastocyst implanted

31. _____ Links the embryo with the placenta

32. _____ Source of early blood cells and of germ cells that will become the gametes

33. _____ Membrane that develops into a fluid-filled sac that surrounds the embryo and later, the fetus

34. _____ Parts of this membrane give rise to the embryo's digestive tube

35. _____ Fluid within this membrane keeps the embryo from drying out, absorbs shocks, and acts as insulation.

36. _____ Produces blood vessels that invade the umbilical cord

37. _____ Through this tissue, the embryo receives nutrients and oxygen from the mother and sends out wastes to the mother's bloodstream in return.

38. _____ The "maternal side" of this structure consists of a layer of endometrial tissue containing arterioles and venules.

39. _____ Small projections of this membrane extend into spaces filled with maternal blood.

40. _____ In addition to oxygen and nutrients, many other substances taken in by the mother—including alcohol, caffeine, drugs, pesticide residues, HIV, and toxins in cigarette smoke—can cross this structure.

17.7. THE SECOND FOUR WEEKS: HUMAN FEATURES APPEAR [pp.340-341]

17.8. DEVELOPMENT OF THE FETUS [pp.342-343]

17.9. BIRTH AND BEYOND [pp.344-345]

Boldfaced Terms

fetus _____

_____

lactation _____

_____

Dichotomous Choice

Select the term in parentheses that best completes the sentence. [pp.340-341]

1. In an embryo's first few weeks of life, its cells begin to (stabilize / specialize).

2. Growth of the (heart / head) surpasses that of any other body region.

3. As the first 8 weeks of development end, the developing baby has features that define it as a(n) (human / vertebrate).

4. As the second half of the first trimester begins, (gonads / chorionic villi) begin to form.

5. The presence of the Y chromosome causes the development of (testes / a penis) which then make hormones that masculinize the embryo.

6. Female gonads develop as a result of (the presence of estrogen / the absence of testosterone).

7. Once the organ systems are formed, the embryo is designated a(n) (human / fetus).

8. At the end of the first trimester, a baby's sex (can / cannot) be determined.

9. After eight weeks of development, the embryo is about (one inch/ six inches) long.

10. The heartbeat can first be heard at the end of the (first/ second) trimester.

Fill-in-the-Blanks [pp.342-343]

When the fetus is three months old, soft hair called the (11) _____ , covers the fetal body. The skin is wrinkled and protected by a thick, cheesy coating called the (12_____ . The (13) _____ trimester of human development extends from the start of the fourth month to the end of the sixth. Facial muscles move, and near the end of the trimester, the mother feels the fetus' arms and legs (14) _____ . Eyelids and eyelashes form.

The (15) _____ trimester extends from the seventh month until birth. Not until the middle of the third trimester can the baby (16) _____ on its own. Babies born before seven months' gestation often suffer from respiratory (17) _____ syndrome. The circulatory system takes a detour on its way to independence. Because the fetus exchanges gases and receives nutrients via the mother's bloodstream prior to birth, the fetal (18) _____ system develops temporary vessels that function until birth. Two (19) _____ arteries within the umbilical cord transport deoxygenated blood and metabolic wastes from the fetus to the (20) _____ . Oxygenated blood, enriched with nutrients, returns from the placenta to the fetus in the umbilical (21) _____ .

Other temporary vessels divert blood past the (22) _____ and (23) _____ . These organs do not develop as rapidly as some others, because (by way of the placenta) the (24) _____ body can perform their functions. Fetal lungs are (25) _____ and do not become functional for gas exchange until the newborn takes its first breaths outside the womb. A little of the blood entering the heart's right (26) _____ flows into the right ventricle and moves on to the lungs but most of it travels through a gap in the interior heart wall called the (27) _____ or into an arterial duct, the ductus arteriosus, that entirely bypasses the nonfunctioning lungs.

Likewise, most blood bypasses the fetal liver because the mother's liver performs most liver functions until (28) _____ . Nutrient-laden blood from the placenta travels through a venous duct past the liver and on to the (29) _____ , which pumps it to body tissues. At birth, blood pressure in the heart's left atrium (30) _____ . This causes a flap of tissue to close off the (31) _____ , which gradually seals. The closure separates the (32) _____ and (33) _____ circuits of blood flow, and the arterial duct collapses. The temporary fetal vessels gradually (34) _____ during the first few weeks after birth.

Labeling

Identify each indicated part of the following illustrations. [p.340]

35. _____   42. _____

36. future _____   43. _____

37. _____   44. _____

38. developing _____   45. future _____

39. _____   46. _____

40. _____   47. _____

41. _____

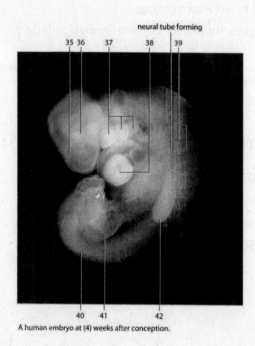

neural tube forming

A human embryo at (4) weeks after conception.

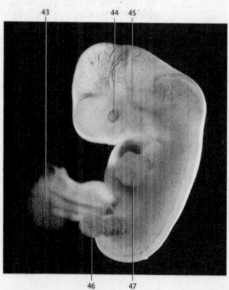

A human embryo at (5–6) weeks after conception.

Development and Aging   273

True-False

If the statement is true, write a "T" in the blank. If the statement is false, make it correct by changing the underlined word(s) and writing the correct word(s) in the answer blank. [pp.344-345]

48. _____ Birth, also called parturition, takes place about 39 weeks after fertilization.

49. _____ "Labor" involves contraction of skeletal muscles in the uterus.

50. _____ Labor is the indirect result of a cascade of hormones from the mother's hypothalamus, pituitary and adrenal glands.

51. _____ In the first stage of labor, the fetus is pushed against its mother's cervix.

52. _____ Usually the amniotic sac ruptures during the second stage of labor.

53. _____ The second stage of labor is the birth of the fetus.

54. _____ Uterine contractions and pushing by the mother move the fetus through the oviduct and out the vaginal canal.

55. _____ The breech position refers to a birth in which the "bottom" rather than the head emerges first.

56. _____ The third stage of labor involves the expulsion of fluid, blood and the endometrium from the mother's body.

57. _____ The navel is the scar left from the attachment point of the umbilical cord.

58. _____ Without the placenta to remove wastes, carbon dioxide builds up in the baby's blood, prompting it to cry.

59. _____ The foramen ovale normally closes within the first month of life.

60. _____ Babies born prematurely have complications because their organs are not developed to point that they can function independently.

61. _____ For the first few days after a baby's birth, glands in a mother's breasts produce a pale fluid called bilirubin.

62. _____ The "let-down" of milk is triggered by a newborn nursing.

63. _____ In response to mechanoreceptor signals, the mother's hypothalamus signals the pituitary to release oxytocin.

64. _____ Oxytocin causes milk let-down as well as contraction of placental muscles that help to shrink the uterus back to normal size.

17.10. DISORDERS: MISCARRIAGE, STILLBIRTHS AND BIRTH DEFECTS [pp.346-347]

17.11. PRENATAL DIAGNOSIS: DETECTING BIRTH DEFECTS [p.348]

17.12. FROM BIRTH TO ADULTHOOD [p.349]

17.13. TIME'S TOLL: EVERYBODY AGES [pp.350-351]

Boldfaced Terms

spina bifida _____

_____

rubella _____

_____

fetal alcohol syndrome_____

_____

chorionic villus sampling _____

_____

amniocentesis _____

_____

preimplantation diagnosis _____

_____

puberty _____

_____

senescence _____

_____

Alzheimer's disease _____

_____

Short Answer

1. Compose a few statements about the critical importance of maternal lifestyle during pregnancy. [pp.346-347]

_____

_____

_____

_____

_____

Matching

Choose the most appropriate answer for each term. [pp.346-347]

2. ____ spina bifida

3. ____ folic acid

4. ____ bacterial infection

5. ____ teratogen

6. ____ viral infection

7. ____ drugs

8. ____ fetal alcohol syndrome

9. ____ cigarette smoke

A. Mother's IgG antibodies protect developing infant from almost all of these

B. Reduces level of vitamin C in blood of pregnant woman and fetus

C. Birth defect in which neural tube doesn't close and separate from ectoderm

D. First trimester developing infant shows great sensitivity to these

E. Small brain and head, facial deformities, poor motor coordination

F. Agent that causes a birth defect

G. Nutrient required for proper development of neural tube

H. Rubella can cause birth defects early in pregnancy

Dichotomous Choice

Circle one of two possible answers given between parentheses in each statement. [p.348]

10. (Amniocentesis / CVS) uses tissue from the chorionic villi of the placenta for prenatal diagnosis of genetic defects.

11. (Amniocentesis / CVS) samples fluid from within the amnion that contains some fetal cells and chemicals used for prenatal diagnosis of genetic defects.

12. Using methods of (CVS / preimplantation diagnosis), an embryo "conceived" by in vitro fertilization is analyzed for genetic defects using recombinant DNA technology.

13. [Ultrasound / Fetoscopy] uses a fiber-optic endoscope uses sound waves to diagnose blood disorders in a fetus.

14. In the period known as [adulthood / adolescence] that follows puberty, an individual becomes physically, mentally and emotionally mature.

15. During the first two weeks after birth, a newborn is referred to as a (neonate / infant).

16. The progressive cellular and bodily deterioration is built into the life cycle of all organisms; the process is called (senescence / adulthood).

17. The period from 2 weeks to about 15 months is (pubescence / infancy).

18. All the developmental stages from 2 weeks after fertilization until the end of the eighth week refers to the (fetus / embryo).

19. All developmental stages from the ninth week until birth refers to the (fetus / embryo).

20. A single cell resulting from fertilization is a (morula / zygote).

Matching

Choose the most appropriate answer for each term. [pp.349-351]

21. _____ aging or senescence

22. _____ cumulative assaults to DNA

23. _____ aging skin, muscles, and the skeleton

24. _____ aging in the cardiovascular and respiratory systems

25. _____ aging of the nervous system and senses

26. _____ aging of the reproductive system and changes in sexuality

27. _____ aging of the immune system

28. _____ aging of the digestive system

29. _____ aging of the endocrine system

30. _____ telomeres

A. Menopause in women; falling levels of testosterone and reduced fertility in men

B. Most hormone levels remain steady with the exception of sex hormones

C. Number of T cells falls, B cells become less active, ability to recognize self markers on body cells declines, more prone to autoimmune diseases

D. Free radical damage of proteins, DNA and other biological molecules; cells lose the capacity for DNA self-repair; mutations accumulate

E. Walls of alveoli break down, enlarged heart, weaker heart muscles, plaques, high blood pressure

F. Numbers of fibroblasts in dermis decreases; elastic fibers replaced with more rigid collagen; less elastic skin; more porous bones; joint breakdown and osteoarthritis

G. Physical and physiological changes beginning at about age 40

H. End caps of chromosomes which are gradually lost after 80 to 90 cell divisions

I. Neurofibrillary tangles, beta amyloid protein fragments, and Alzheimer's diseases (AD)

J. Digestive glands in mucous membranes deteriorate, fewer digestive enzymes are secreted

Short Answer

31. Describe the physical changes that occur in Alzheimer's disease. [p.351] _____

_____

_____

_____

32. What gene is involved in inherited Alzheimer's disease? What type of Alzheimer's is linked to this gene?[p.351] _

_____

_____

_____

33. Why do people after age 60 begin to have "senior moments?" [p.351] _____

_____

_____

_____

SELF-QUIZ

Multiple Choice

___ 1. As it travels down the oviduct, the zygote is subdivided into a multicellular embryo through a process known as _____. [p.332]

    a. meiosis
    b. parthenogenesis
    c. embryonic induction
    d. cleavage
    e. invagination

___ 2. In the uterus, the morula is transformed into a(n) _____. [p.334]

    a. zygote
    b. blastocyst
    c. gastrula
    d. third germ layer
    e. organ

___ 3. Muscles differentiate from _____ tissue. [p.332]

    a. ectoderm
    b. mesoderm
    c. endoderm
    d. parthenogenetic
    e. yolk

___ 4. The nervous system differentiates from _____ tissue. [p.332]

    a. ectoderm
    b. mesoderm
    c. endoderm
    d. parthenogenetic
    e. yolk

___ 5. The process by which body tissues and organs form is _____. [p.332]

    a. gastrulation
    b. morphogenesis
    c. cleavage
    d. gamete formation
    e. fertilization

___ 6. Genetically programmed cell death is _____. [p.337]

    a. diaphoresis
    b. morphogenesis
    c. senescence
    d. apoptosis
    e. amniocentesis

Matching

7. _____ HCG [p.335]

8. _____ second trimester [p.342]

9. _____ umbilical cord [p.338, p.344]

10. _____ first trimester [pp.340-341]

11. _____ conjoined twins [p.335]

12. _____ fetus [p.341]

13. _____ amnion [p.338]

14. _____ placenta [p.338]

15. _____ chorion [p.338]

16. _____ lactation [p.345]

17. _____ third trimester [p.342]

18. _____ implantation [p.334]

19. _____ FAS [p.347]

20. _____ puberty [p.349]

A. Outermost extraembryonic membrane, secretes HCG to maintain the uterine lining for three months

B. Period of development extending from the start of the fourth month to the end of the sixth

C. Blastocyst adheres to the uterine lining; cells invade maternal tissues

D. Fluid-filled sac immediately surrounding the embryo

E. Period of development extending from the seventh month until birth

F. Hormone secreted by the blastocyst; stimulates corpus luteum to continue estrogen and progesterone secretion to maintain uterine lining

G. Disease common to two-thirds of pregnant alcoholics

H. Occurs when an embryo partially splits after day 12 of development

I. Period of development from fertilization to the end of the third month

J. Blood vessels of this structure are used to transport oxygen and nutrients for the embryo

K. Arrival of sexual maturity and functioning of the reproductive organs

L. Embryo becomes this after organ system formation is completed, beginning about the ninth week

M. Structure connecting embryo or fetus with the placenta

N. Period during which hormone-primed glands produce milk

CHAPTER OBJECTIVES / REVIEW QUESTIONS

1. When the three germ layers have formed and different genetic instructions begin to operate in different groups of cells, the process of _____ begins. [p.332]

2. Name each of the three germ layers, and generally list the organs formed from each. [p.332]

3. From which part of the blastocyst does the embryo develop? [p.334]

4. Describe the processes and products of morphogenesis. [p.332-333]

5. Define apoptosis, give an example of the results of this process and explain what might happen if this process does not occur. [p.337]

6. Define amniocentesis and explain what it is used for. [p.348]

7. Define fetoscopy. [p.348]

8. A(n) _____ consists of a surface layer of cells called the *trophoblast* and a clump of cells to one side called the *inner cell mass*. [p.334]

9. How much time passes between fertilization and implantation? [p.334]

10. Describe the process of implantation. [p.334-335]

11. The _____, will give rise to the brain and spinal cord. [p.336]

12 Name the four membranes that form around the early embryo, and give the major functions of each. [p.338]

13 The embryo and mother exchange substances through the _____. [p.339]

14 At what point in the development process does the embryo begin to be referred to as a fetus? [p.341]

15 Explain why the mother must be particularly careful of her diet, health habits, and lifestyle during the first trimester after fertilization. [pp.346-347]

INTEGRATING AND APPLYING KEY CONCEPTS

Define the term teratogen. Give three examples of teratogens and describe their affects on the developing fetus. Estimate at which point in time during the fetus' development these teratogens might exert their effects.

18

CELL REPRODUCTION

INTRODUCTION

Cell reproduction occurs in two different ways. One involves mitosis and the other meiosis. As you study this chapter, it is critical to distinguish between the two in both purpose and sequence of events. Remember that mitosis is used for somatic cells and to make copies with the same chromosome number while meiosis is used to make gametes (sperm or eggs). Follow the progress of chromosomes through both types of division. Mastering the vocabulary describing chromosomes should be your first goal. As you do, it will help to keep in mind what the end result of the division will produce so that the events leading to it will make sense.

FOCAL POINTS

- Figure 18.3 [p.358] is a diagram of the cell cycle. It is important to recognize that all human cells begin their existence in G_1 of this cycle. All cells destined to divide complete all of Interphase. The two types of cell division become different after interphase.
- Figure 18.6 [p.361] illustrates the stages of mitosis.
- Figures 18.9 and 18.10 [p.365] illustrate the formation of human sperm and eggs. It is helpful in studying meiosis to relate the process to the production of these gametes.
- Figures 18.12 [p.368] and 18.13 [p.369] show the events of crossing over and random alignment. Since meiosis is the source of much of the tremendous variation associated with sexual reproduction, understanding these events is extremely important, and will make the study of inheritance in future chapters much simpler.
- Figure 18.14 [pp.370-371] compares mitosis and meiosis. This figure is helpful in recalling the stages of either process as well as in comparing the two. When you have completely mastered the two processes, you should be able to draw the events shown in this figure yourself.

INTERACTIVE EXERCISES

CHAPTER INTRODUCTION [P. 355]

18.1. REPRODUCTION: CONTINUING THE LIFE CYCLE [PP.356-357]

Boldfaced Terms

reproduction _____

_____

life-cycle _____

_____

chromosomes _____

_____

chromosome number _____

_____

diploid _____

_____

autosomes _____

_____

sex chromosomes _____

_____

homologous chromosomes _____

_____

haploid _____

_____

Matching

Choose the most appropriate description for each term. [pp. 356-357]

1. ____ autosome
2. ____ diploid
3. ____ homologous chromosomes
4. ____ haploid
5. ____ life cycle
6. ____ sex chromosome
7. ____ chromosome number
8. ____ reproduction
9. ____ chromosome
10. ____ somatic

A. When a parent cell produces a new generation of cells, or when parents of a species produce a new individual

B. Indicates the sum of all chromosomes normally present in a cell

C. Refers to a cell containing only one of each chromosome - n

D. A pair of physically similar chromosomes (one from each parent) with genetic instructions for the same traits

E. Single DNA molecule combined with protein, location of genes

F. Refers to body cells

G. A recurring series of events in which individuals grow, develop, maintain themselves, and reproduce according to instructions encoded in DNA

H. Refers to a cell possessing two of each type of chromosome - $2n$

I. X or Y

J. Chromosome other than a sex chromosome

Dichotomous Choice

Circle one of the two possible answers given between parentheses in each statement. [p.356]

11. A chromosome consists of DNA attached to proteins called (spindles / histones).

12. A unit of DNA complexed with a histone is a (nucleosome / centromere).

13. As a cell nucleus prepares for division, each chromosome (condenses / elongates).

18.2. OVERVIEW OF THE CELL CYCLE AND CELL DIVISION [P.358-359]

18.3. THE FOUR STAGES OF MITOSIS [P.360-361]

Boldfaced Terms

mitosis _____

_____

meiosis _____

_____

sister chromatids _____

_____

centromere _____

_____

cell cycle _____

_____

G1 _____

_____

S _____

_____

G2 _____

_____

M _____

_____

chromatin _____

_____

prophase _____

_____

metaphase _____

_____

anaphase _____

_____

telophase _____

_____

spindle _____

_____

Labeling-Matching

Identify each of the following mitotic stages by entering the correct stage in the blank beneath the sketch. Select from *transition to metaphase, metaphase, early prophase, prophase, telophase,* and *anaphase*. Complete the exercise by matching and entering the letter of the correct phase description in the parentheses after each label. [p. 361]

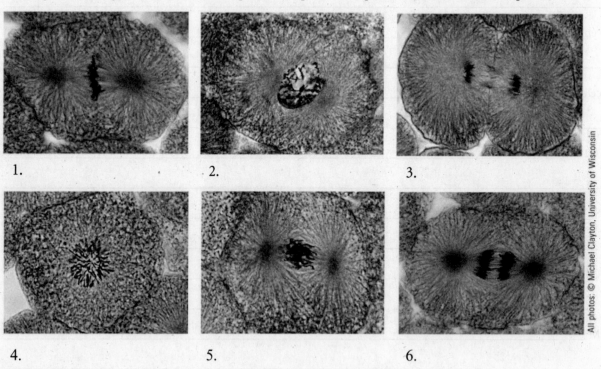

1.　　　　　　　　2.　　　　　　　　3.

4.　　　　　　　　5.　　　　　　　　6.

All photos: © Michael Clayton, University of Wisconsin

A. Attachment between two sister chromatids of each chromosome breaks; the two are now chromosomes in their own right; they move to opposite spindles.

B. Microtubules that form the spindle apparatus enter the nuclear region; microtubules become attached to the sister chromatids of each chromosome.

C. The DNA and its associated proteins start to condense into the threadlike chromosome form.

D. Chromosomes are now fully condensed and lined up at the equator of the spindle.

E. Chromosomes continue to condense; new microtubules are assembled, and they move one of two centrioles toward the opposite end of the cell: the nuclear envelope breaks up.

F. The chromsomes reach the spindle poles and a nuclear envelope forms around each cluster.

True/False

Write True on the blank line if the statement is true or False on the blank line if the statement is false. [p. 359]

7. _____ The function of mitosis is for the production of gametes.

8. _____ The effect of mitosis is the creation of two haploid daughter cells.

Fill-in-the-Blanks [pp. 358-361]

(9) _____, the first stage of mitosis, is evident when chromosomes become visible in the light microscope as threadlike forms. Each chromosome was duplicated earlier, during (10) _____. Each already consists of two (11) _____ joined together at the (12) _____. Early in prophase, the sister chromatids of each (13) _____ twist and fold into a more compact form. By late prophase, all the chromosomes will be (14) _____ into thick, rod-shaped forms. Cytoplasmic microtubules break apart into their tubulin subunits. The subunits reassemble near the nucleus, as new (15) _____. Many cells have two barrel-shaped (16) _____ that were duplicated during interphase. In prophase, the cell has (17) _____ of centrioles. Microtubules begin moving one pair to the opposite pole of the developing spindle.

(18) _____ is a time during which the nuclear envelope breaks up completely, into numerous tiny, flattened vesicles. The (19) _____ begin interacting with the microtubules. When each chromosome is harnessed by microtubules from both (20) _____, a two-way pull orients the two sister (21) _____ of a chromosome toward opposite poles. When all the duplicated chromosomes are aligned midway between the poles of a completed spindle, the cell is in (22) _____.

During (23) _____, the two sister chromatids of each chromosome separate and move to opposite poles by two mechanisms: microtubules attached to the centromere regions shorten, pulling the chromosomes to the poles, and the spindle elongates when overlapping microtubules ratchet past each other and push the spindle poles farther apart. Once separated from its sister, each (24) _____ now becomes a chromosome in its own right.

Cell Reproduction   **285**

(25) _____ begins when two clusters of chromosomes arrive at opposite spindle poles.

Chromosomes are no longer harnessed to microtubules, and they return to a more threadlike form. Bit by bit a new

(26) _____ forms around each cluster of chromosomes separating it from the cytoplasm. With mitosis,

each new nucleus has the same chromosome (27) _____ as the parent nucleus. Once the two nuclei

form, (28) _____ is over—and so is (29) _____.

Labeling

Identify the stage in the cell cycle indicated by each number. [p.358]

30. (1) _____

31. (5) _____

32. (2) _____

33. (3) _____

34. (4) _____

35. (6) _____

36. (7) _____

37. (8) _____

38. (9) _____

39. (10) _____

© Cengage Learning. All Rights Reserved.

Matching

40. Link each of the following time spans to the parts of the cell cycle by placing numbers from the diagram above into the blanks below. Answers may be used more than once. [p.358]

_____ Period after DNA replication, cell prepares to divide

_____ Chromosomes are sorted into two sets, followed by cytoplasmic division

_____ DNA replication occurs

_____ Period of cell growth before DNA replication

_____ Longest phase of the cell cycle

_____ Period of cytoplasmic division

_____ Period that includes G_1, S, and G_2

_____ 41. The "pinched in" region of a replicated chromosome is called a:

       A. G1 growth ring.

       B. chromatid.

       C. centromere.

       D. microfiber attachment ring.

_____ 42. DNA is duplicated during which of the following stages of mitosis:

       A. G2.

       B. G1.

       C. synthesis.

       D. B12.

       E. none of these.

_____ 43. Sister chromatids:

       A. are made of two different chromosomes.

       B. are two copies of the same chromosome.

       C. are attracted to one another due to magnetic charges at their core.

       D. are only produced in mammals and higher animals.

       E. none of these statements is true.

Dichotomous choice

Circle one of the two possible answers given between parentheses in each statement.   [p.356]

44. A cell that has two of each type of chromosome is called a (haploid / diploid) cell.

45. A duplicated chromosome consists of two copies called (sister / twin) chromatids.

46. The "pinched in" region of a replicated chromosome is called a (chromatid / centromere) and is where (microtubules / histones) attach.

47. The (centromere / spindle) consists of two sets of microtubules extending from the two poles of the cell.

48. The poles from which microtubules extend are (centrioles / centromeres).

Choice

For questions 49–54, choose from the following: [pp.358-359]

    a. mitosis   b. meiosis

49. _____ Reduces chromosome number from diploid to haploid

50. _____ Involves one round of chromosome partitioning and cytokinesis

51. _____ Starts with a diploid cell and ends with two diploid cells

52. _____ Occurs in germ cells of testes and ovaries

53. _____ Involves two rounds of chromosome partitioning and cytokinesis

54. _____ Starts with a diploid cell and produces four haploid daughter cells

18.4. HOW THE CYTOPLASM DIVIDES [PP.362]

18.5. CONCERNS AND CONTROVERSIES OVER IRRADIATION [P.363]

Boldfaced Terms

cytokinesis _____

_____

cleavage furrow _____

_____

Matching

Match the descriptions of cytokinesis with the four diagrams. [p.362]

1. ____

2. ____

3. ____

4. ____

A. Shrinkage of the microfilament ring pulls the cell surface inward resulting in a cleavage furrow.

B. Mitosis is complete and the spindle is disassembling.

C. Contractions continue; the cell is pinched in two.

D. A ring of microfilaments attached to the plasma membrane contracts.

Short Answer

5. List some common natural sources of ionizing radiation. [p.363] _____

_____

_____

_____

6. Summarize the effects of ionizing radiation on cells. [p.363] _____

_____

_____

_____

7. Name some uses of irradiation in medicine and the food industry. [363] _____

_____

_____

_____

18.6. MEIOSIS: THE BEGINNINGS OF EGGS AND SPERM [PP.364-365]

18.7. THE STAGES OF MEIOSIS [PP.366-367]

Boldfaced Terms

spermatogenesis_____

_____

oogenesis _____

_____

Comparison

Fill in the chart below in order to compare and contrast mitosis and meiosis. Assume that both processes begin with diploid (2n) cells. [p.364]

	Mitosis	*Meiosis*
1. # of times DNA replication occurs	1	
2. # of times cell division occurs	1	
3. # of resulting cells	2	
4. ploidy (*n* or *2n*) of final daughter cells	*2n*	
5. purpose in human life cycle	growth, repair	

Comparison

Fill in the chart below in order to compare and contrast the two stages of meiosis. [p.364]

	Meiosis I	*Meiosis II*
6. homologous chromosomes are separated		
7. begins with one *2n* cell, ends with two *n* cells		
8. chromatids are pulled apart		
9. begins with two *n* cells, ends with four *n* cells		
10. directly preceded by DNA replication		

Matching

To review the major events of spermatogenesis, match the following written descriptions with the appropriate sketch. Place the correct number of the sketch on the blank line. To simplify the process, assume that the cell in this model initially has only two pairs of homologous chromosomes (two from a paternal source and two from a maternal source) and that crossing over does not occur. Complete the exercise by indicating diploid ($2n$) or haploid (n) chromosome number of the cell chosen in the parentheses following each blank. [p.365]

11. _____ ( ) Spermatids mature as sperm.

12. _____ ( ) A male germ cell develops into a primary spermatocyte as it replicates its DNA.

13. _____ ( ) Meiosis I in the primary spermatocyte results in two secondary spermatocytes.

14. _____ ( ) Four spermatids form when the secondary spermatocytes undergo meiosis II.

15. _____ ( ) Male germ cell.

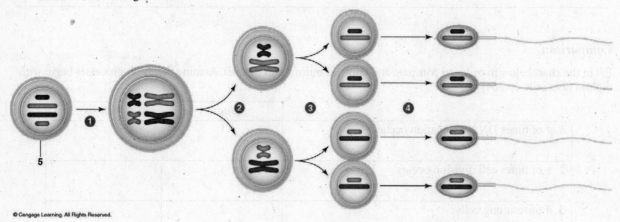

Sequence

Arrange the following entities in correct order of development, entering a 1 by the stage that appears first and a 5 by the stage that completes the process of spermatogenesis. Complete the exercise by indicating if each cell is n or $2n$ in the parentheses following each blank. [pp.364–365]

16. _____ ( ) primary spermatocyte

17. _____ ( ) sperm

18. _____ ( ) spermatid

19. _____ ( ) spermatogonium

20. _____ ( ) secondary spermatocyte

Short Answer

21. List the major differences between oogenesis and spermatogenesis. [pp.364-365] _____

_____

_____

_____

_____

Labeling-Matching

This section will help you review the details of meiosis I and II. Identify each of the following meiotic stages by entering the correct stage of either meiosis I or meiosis II in the blank beneath the sketch. Choose from *prophase I, metaphase I, anaphase I, telophase I, prophase II, metaphase II, anaphase II,* and *telophase II*. Complete the exercise by matching and entering the letter of the correct phase description in the parentheses following each label. [pp.366–367]

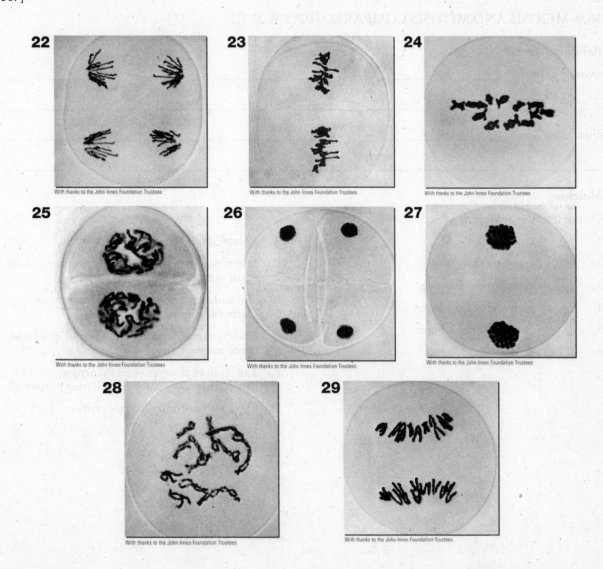

A. The spindle is now fully formed; all chromosomes are positioned midway between the poles of one cell.

B. Centrioles have moved to opposite poles in each of two cells, a spindle forms, and microtubules attach the duplicated chromosomes to the spindle and begin moving them.

C. Four daughter nuclei form; when the cytoplasm divides, each new cell has a haploid number of chromosomes, all in the unduplicated state; one or all cells may develop into gametes.

D. In one cell, each duplicated chromosome is pulled away from its homologue; the two are moved to opposite spindle poles.

E. Each chromosome is drawn up close to its homologue; crossing over and swapping of segments (genetic recombination) occurs.

F. All the chromosomes are now aligned at the spindle's equator; this is occurring in two haploid cells.

G. Two haploid cells form, but chromosomes are still in the duplicated state.

H. Chromatids of each chromosome separate; former "sister chromatids" are now chromosomes in their own right and are moved to opposite poles.

18.8. HOW MEIOSIS PRODUCES NEW COMBINATIONS OF GENES [PP.368-369]

18.9. MEIOSIS AND MITOSIS COMPARED [PP.370-371]

Boldfaced Terms

crossing over _____

_____

disjunction _____

_____

Matching

Choose the most appropriate description for each term. [pp. 368-369]

1. ____ crossing over
2. ____ nonsister chromatids
3. ____ genetic recombination
4. ____ maternal chromosomes
5. ____ paternal chromosomes
6. ____ disjunction

A. Chromosomes inherited from mother

B. Interaction between two nonsister chromatids of a pair of homologues during prophase I

C. The separation of each homologue from its partner during anaphase I

D. One chromatid from each of a homologous pair of chromosomes

E. The result of chromosomes exchanging pieces between nonsister chromatids during prophase I

F. Chromosomes inherited from father

Complete the Table

7. Complete the following table by entering the word *mitosis* or *meiosis* in the blank adjacent to the statement describing one of the two processes. [pp.370-371]

Type of cell division	Description
a.	Involves one division
b.	Daughter cells are haploid
c.	Involves two divisions
d	Daughter cells have one chromosome from each homologous pair
e.	Daughter cells have the diploid chromosome number
f.	Completed when four daughter cells are formed

Short Answer

8. List and briefly describe the three sources of genetic variation resulting from sexual reproduction (meiosis and fertilization) in animals. [p.368]_____

_____

_____

_____

_____

Matching

The cell model used in this exercise has two pairs of homologous chromosomes, one long pair and one short pair. Match the descriptions to the letters of the sketches below. [pp.370-371]

9. ___ Metaphase of mitosis

10. ___ Metaphase I of meiosis

11. ___ Anaphase II of meiosis

12. ___ Anaphase I of meiosis

13. ___ Prophase of mitosis

14. ___ Prophase I of meiosis

15. ___ Prophase II of meiosis

16. ___ Telophase of mitosis

17. ___ Telophase I of meiosis

18. ___ Telophase II of meiosis

19. ___ Metaphase II of meiosis

20. ___ Anaphase of mitosis

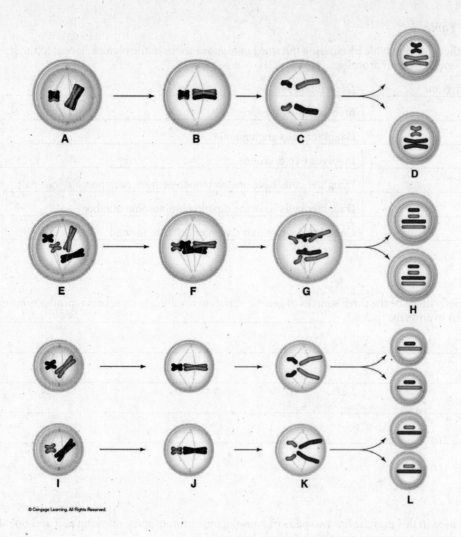

© Cengage Learning. All Rights Reserved.

SELF-QUIZ

____ 1. The DNA copying process occurs
_____. [p.358]

    a. during the S phase of interphase
    b. during the G_2 phase of interphase
    c. during prophase of mitosis
    d. during the G_1 phase of interphase

____ 2. In eukaryotic cells, which of the following
occurs during mitosis? [p.360]

    a. cytoplasmic division
    b. replication of DNA
    c. a long growth period
    d. sorting of chromosomes into two nuclei

____ 3. Being "diploid" means having _____.
[p.356]

    a. two chromosomes of each type in
somatic cells
    b. twice the parental chromosome number
    c. half the parental chromosome number
    d. one chromosome of each type in
somatic cells

____ 4. Somatic cells are _____ cells; germ cells
are _____ cells. [p.356]

    a. meiotic; body
    b. body; body
    c. meiotic; meiotic
    d. body; meiotic

5. If a parent cell has sixteen chromosomes and undergoes mitosis, the resulting cells will have _____ chromosomes. [p.356-357]

   a. sixty-four
   b. thirty-two
   c. sixteen
   d. eight
   e. four

6. The correct order of the stages of mitosis is _____. [pp.360-361]

   a. prophase, metaphase, telophase, anaphase
   b. telophase, anaphase, metaphase, prophase
   c. telophase, prophase, metaphase, anaphase
   d. anaphase, prophase, telophase, metaphase
   e. prophase, metaphase, anaphase, telophase

7. Which of the following does not occur in prophase I of meiosis? [pp.366]

   a. a cytoplasmic division
   b. a cluster of four chromatids (2 chromosomes)
   c. homologues pairing tightly
   d. crossing over

8. Crossing over is one of the most important events in meiosis because _____. [pp.368]

   a. it produces new combinations of alleles on chromosomes
   b. homologous chromosomes must be separated into different daughter cells
   c. the number of chromosomes allotted to each daughter cell must be halved
   d. homologous chromatids must be separated into different daughter cells

9. If a parent cell has sixteen chromosomes and undergoes meiosis, the resulting cells will have _____ chromosomes. [p.364]

   a. sixty-four
   b. thirty-two
   c. sixteen
   d. eight
   e. four

10. Which of the following does not increase genetic variation? [p.368]

    a. crossing over
    b. random fertilization
    c. prophase of mitosis
    d. random homologue alignments at metaphase I

11. Which of the following is the most correct sequence of events in human life cycles? [p.368]

    a. meiosis → fertilization → gametes → diploid organism
    b. diploid organism → meiosis → gametes → fertilization
    c. fertilization → gametes → diploid organism → meiosis
    d. diploid organism → fertilization → meiosis → gametes

12. During cell division, when a cleavage furrow forms, the progression to the actual formation of two new cells is brought about by microfilaments made of the protein _____ pinching the cell membrane in two. [pp.362]

    a. myosin
    b. titin
    c. actin
    d. troponin

13. The cell in the diagram below is a diploid cell that has two pairs of chromosomes. From the number and pattern of chromosomes, the cell could be in _____. [p.361]

    a. the first division of meiosis
    b. the second division of meiosis
    c. mitosis
    d. either a or c

CHAPTER OBJECTIVES / REVIEW QUESTIONS

1. Distinguish between somatic cells and germ cells as to their location and function. [pp.358-359]

2. *Mitosis* and *meiosis* refer to the sorting and packaging of the cell's _____. [p.358-359]

3. State the criteria that two chromosomes must meet to be called "homologous chromosomes." [p.357]

4. Distinguish between autosomes and sex chromosomes. How many of each does a diploid human cell contain? A haploid human cell? [pp.356-357]

5. The eukaryotic chromosome is composed of _____ and _____. [p.356]

6. The two attached copies of a duplicated chromosome are known as sister _____. [p.358]

7. List, in order, the stages and the various activities occurring during the eukaryotic cell cycle. [p.358]

8. Describe the function of the portion of a chromosome known as a *centromere*. [p.358]

9. While diploid germ cells are in the S phase of interphase, each chromosome is _____; each chromosome is then composed of two sister _____. [p.358-359]

10. Describe what occurs in each of the four stages of mitosis. Starting with a cell with two chromosomes in G_1 of mitosis, draw the cell at the end of interphase and at each mitotic stage, ending with two daughter cells in G_1 of the next cell cycle. [pp.360-361]

11. Tell what happens in animal cell cytokinesis. During which mitotic stage does this occur? [p.362]

12. What are the two different divisions of meiosis called? [p.364]

13. Describe spermatogenesis in human males. [p.364]

14. Describe oogenesis in human females. [pp.364]

15. If the diploid chromosome number for humans is 46, the haploid number is _____. [p.357]

16. In prophase of meiosis I, homologous chromosomes exchange corresponding genes in a process known as _____. How is this process beneficial? [p.368]

17. What occurs during metaphase I that also produces genetic variation? [p.368-369]

INTEGRATING AND APPLYING KEY CONCEPTS

1) Runaway cell division is characteristic of cancer. Imagine the various points of the mitotic process that might be sabotaged in cancerous cells in order to halt their multiplication. Then try to imagine how one might discriminate between cancerous and normal cells in order to guide those methods of sabotage most effective in combating cancer.

2) Colchicine is a chemical used to treat gout. Unfortunately, its mechanism of action includes interference with mitosis. Explain how it does this and discuss some of the possible consequences of this activity.

19

OBSERVABLE PATTERNS OF INHERITANCE

INTRODUCTION TO GENETICS

While it is fairly simple to predict the inheritance of some traits, many patterns of inheritance are complex. This chapter starts by defining essential terms and explaining the prediction of single traits or pairs of traits that follow the rule of dominance. Before proceeding to more complex inheritance mechanisms, you should be comfortable with analyzing these simple crosses and predicting what the offspring will be like. At that point, you are ready to deal with the interesting variety of inheritance patterns. This information can be applied to plant and animal breeding endeavors and to studies of genetic disorders.

FOCAL POINTS

- Figure 19.1 [p.376] explains that genes encode different traits and frequently occur in pairs called alleles.
- Figures 19.6 and 19.7 [p.379] demonstrate the steps used to predict the outcome of a genetic cross by the Punnett square method as well as by mathematically calculating probabilities. The former method is based on the latter. Whichever you use, you should get the same answer.
- Figures 19.9 and 19.10 [p.381] extends the use of Punnett squares and probability equations to predicting the inheritance pattern of two traits, rather than just one.

INTERACTIVE EXERCISES

CHAPTER INTRODUCTION [P. 375]

19.1. BASIC CONCEPTS OF HEREDITY [P.376]

19.2. ONE CHROMOSOME, ONE COPY OF A GENE [P.377]

19.3. GENETIC TOOLS: TESTCROSSES AND PROBABILITY [PP.378-379]

Boldfaced Terms

genes _____

_____

locus_____

_____

allele_____

_____

homozygous_____

_____

heterozygous_____

_____

genotype_____

_____

phenotype_____

_____

segregation_____

_____

probability_____

_____

Punnett square_____

_____

Matching

Choose the most appropriate description for each term. [p.376]

1. ____ alleles
2. ____ heterozygous
3. ____ dominant allele
4. ____ phenotype
5. ____ genes
6. ____ recessive allele
7. ____ homozygous
8. ____ diploid cell
9. ____ gene locus
10. ____ genotype

A. The different versions of a gene

B. Location of a specific gene on a specific chromosome

C. Describes an individual having a pair of different alleles

D. Gene whose effect may be masked by its partner allele

E. Refers to an individual's actual physical or functional traits

F. Refers to the alleles inherited by an individual

G. Manipulation of genes in the eggs or sperm so that a child would have traits desirable to parents

H. Describes an individual with two identical alleles

I. Units of information about specific traits; passed from parents to offspring

J. Has two copies of each gene, on pairs of homologous chromosomes

K. Gene whose effect may mask the effect of its partner

Dichotomous Choice

Circle one of two possibilities given between parentheses in each statement. [pp. 376-379]

11. A genetic cross involving only one trait is a (homologous / monohybrid) cross.

12. When eggs or sperm are produced, each receives one allele for individual traits because only one copy of the _____ is obtained (chromosome / gamete).

13. The separation of homologous chromosomes into different gametes is called (segregation / differentiation).

14. Gametes are always (haploid / diploid).

15. If a parental generation cross is CC x cc, half of the grandchildren's generation will have the genotype (CC / Cc).

16. A (Probability / Punnett) square is a convenient way of determining the probable outcome of a cross.

17. The number between zero and one that expresses the likelihood of an event is (ratio / probability).

18. The outcomes predicted by probability (have to / don't have to) turn up in an individual family.

19. If a ¼ probability for the inheritance of a given phenotype is calculated for the first child in a family, this probability (does / doesn't) change for the second child.

20. A testcross is used to learn the (genotype / phenotype) of an organism.

21. In a testcross, in order to determine if an organism with a dominant genotype is homozygous or heterozygous, it is crossed to an individual that is homozygous (dominant / recessive).

Sequence

Place the steps for completing a Punnett square problem in the correct sequence by placing the letter of the first step in the first blank, and so on. [p.378]

22. ____

23. ____

24. ____

25. ____

a. Determine the genotypes of the parents and the possible gametes that each can make.

b. Draw a square with each possible female gamete at the top of a column, and each possible male gamete beside a row.

c. Find the answer to the problem by presenting the genotypes in the Punnett square as a ratio, and/or interpreting the resulting phenotypic ratio.

d. Fill in the Punnett square by indicating genotypes of possible offspring at the intersection of each column and row.

Problems

26. In humans, chin fissure (represented by *C*) is dominant over smooth chin (represented by *c*). A woman who is heterozygous for this trait marries a man with a smooth chin. [pp.377-379]

   a. What two types of gametes would the woman make?_____

   b. What one type of gamete would the man make? _____

   c. Place these on the appropriate sides of the Punnett square below.

   d. Fill in the four squares of the Punnett square. What ratio of genotypes would you predict in the offspring? _____

   e. What ratio of phenotypes would you predict for the offspring? _____

   f. If the couple has one child with a smooth chin, what is the probability of their second child having a chin fissure? _____

27. The problem above can also be solved by using simple probability equations. The likelihood of the woman producing either *C* or *c* gametes is ½. The probability of the man producing c gametes is 1. [pp.378-379]

    a. Show the equation and solution for predicting the probability of this couple having a *Cc* child.

        _____

    b. Show the equation and solution for predicting the probability of this couple having a *cc* child.

        _____

    c. What fraction of the offspring is predicted to have a chin fissure? What fraction is predicted to have a smooth chin? _____

28. In humans, the gene for the ability to taste phenyl-thio-carbamine (PTC), *A*, is dominant; its allele for the inability to taste PTC is *a*. Two parents, one a taster and one a nontaster, have two children. One child is a taster, and one child is a nontaster. With the principles behind a testcross in mind, determine the genotype of the parent who is a taster. State the reasons for your answer. [pp.378-379]

    _____

    _____

    _____

29. In humans, a dominant gene *C* results in chin fissure, and its allele *c* codes for smooth chin. A man who lacks chin fissure had a mother who exhibited chin fissure and a father who lacked chin fissure. The man is married to a woman with chin fissure whose mother lacked chin fissure but whose father had chin fissure. The man and woman are expecting their first child. State the probability that the child will have a smooth chin or a chin fissure. [pp.378-379]

    _____

    _____

    _____

30. Albinos cannot form the pigments that normally produce skin, hair, and eye color, so albinos have white hair and pink eyes and skin (because the blood shows through). To be an albino, one must be homozygous recessive (*aa*) for the pair of genes that codes for the key enzyme in pigment production. Suppose a woman of normal pigmentation (*A*) with an albino mother marries an albino man. State the kinds of pigmentation possible for this couple's children, and specify the ratio of each kind of child the couple is likely to have. Show the genotype(s) and state the phenotype(s). [pp.378-379]

    _____

    _____

    _____

Multiple Choice

____ 31. An individual with two identical alleles is said to be: [p. 376]

        A. Heterozygous
        B. Homozygous
        C. Allelically challenged
        D. Triallelic
        E. None of these

___ 32. The effect of _____ alleles can be masked by another allele. [376]
   A. dominant
   B. recessive
   C. genotype
   D. phenotype
   E. none of these

___ 33. Breeding experiments between different individuals of the same species can reveal genetic information and is called: [p.379]
   A. Testcross
   B. Genetic analysis
   C. Repopulation
   D. Recessive detection
   E. All of these

19.4. HOW GENES FOR DIFFERENT TRAITS ARE SORTED INTO GAMETES [PP.380-381]

Boldfaced Terms

independent assortment _____

_____

Matching

Choose the most appropriate description for each term. [p.380-381]

1. ____ independent assortment

2. ____ linked genes

3. ____ dihybrid cross

A. Follows the inheritance of two traits

B. Genes so close together on a chromosome that they do not show independent assortment

C. Alleles for different traits are sorted into gametes independently of one another.

Problems

When working genetics problems dealing with two gene pairs, one can visualize the independent assortment of gene pairs located on nonhomologous chromosomes into gametes by use of a fork-line device. Assume that in humans, pigmented eyes (*B*) are dominant (an eye color other than blue) over blue (*b*), and right-handedness (*R*) is dominant over left-handedness (*r*). To learn to solve a problem, cross the parents *BbRr* × *BbRr*. A sixteen-block Punnett square is required, with gametes from each parent arrayed on two sides of the Punnett square (see Figure 19.9 in the text). The gametes receive genes through independent assortment using a fork-line method. [pp.380-381]

4. Calculate the products of the crosses shown in this Punnett square; combine these haploid gametes to form diploid zygotes within the squares. After filling in the square, enter the probability ratios derived within the Punnett square for the phenotypes listed beside the blanks. [pp.380-381]

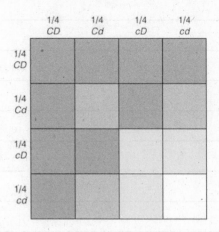

a. _____ chin fissure, dimples

b. _____ chin fissure, no dimples

c. _____ smooth chin, dimples

d. _____ smooth chin, no dimples

5. The problem above can also be worked using probability equations. To do this, the two monohybrid situations within the dihybrid problem are separated. Then, multiplication is used to predict phenotypes for the next generation. See figure 19.10 in text. [p.381]

a. _____ For the trait of chin fissure, what is the probability of the cross Cc x Cc producing offspring with the phenotype of a chin fissure?

b. _____ For the trait of chin fissure, what is the probability of the cross Cc x Cc producing offspring with the phenotype of smooth chin?

c. _____ For the cheek dimples, what is the probability of the cross Dd x Dd producing offspring with the cheek dimple phenotype?

d. _____ For the trait of handedness, what is the probability of the cross Dd x Dd producing offspring with the no dimples phenotype?

e. Using multiplication, determine the probability of the offspring having pigmented eyes and being right-handed. Show the equation.

_____ x _____ = _____

f. What is the probability of the offspring having a chin fissure and no cheek dimples? Show the equation.

_____ x _____ = _____

g. What is the probability of the offspring having a smooth chin and cheek dimples? Show the equation.

_____ x _____ = _____

h. What is the probability of the offspring having a smooth chin and no cheek dimples? Show the equation.

_____ x _____ = _____

19.5. SINGLE GENES, VARYING EFFECTS [PP.382-383]

19.6. OTHER GENE EFFECTS AND INTERACTIONS [PP.384-385]

Boldfaced Terms

pleiotropy _____

_____

sickle-cell anemia _____

_____

codominance _____

_____

multiple allele system _____

_____

penetrance _____

_____

polygenic traits _____

_____

continuous variation _____

_____

multifactorial trait _____

_____

Complete the Table

1. Complete the following table by supplying the principle of inheritance illustrated by each example. Choose from *pleiotropy*, *polygenic inheritance*, *multiple alleles*, and *codominance*.

Type of Inheritance	Example
a.	Expression of a gene affects two or more traits [p.382]
b.	Blood type involves a gene system of three alleles, I^A, I^B, and i. [p.383]
c.	I^A and I^B alleles for blood type are both expressed in an individual [p.383]
d.	Human eye and skin color show continuous variation within a population. [p.384]

Short Answer

2. Distinguish between sickle cell anemia and sickle cell trait, including both genotypic and phenotypic differences. [pp.382-383] _____

_____

_____

_____

3. Why would doctors suggest the addition of the food additive butyrate to the diet of an infant who tests positive for sickle-cell anemia? [p.383] _____

_____

_____

_____

Choice

For questions 4–13, choose from the following concepts concerned with variable gene expression: [pp.382-384]

a. codominance   b. pleiotropy   c. complete penetrance   d. incomplete penetrance
e. variable expressivity   f. polygenic traits   g. sickle cell anemia

4. _____ Continuous variation is seen when there is a range of differences for a trait in a population.

5. _____ Some people who inherit the dominant allele for polydactyly have the normal number of digits, and others have more.

6. _____ This form of inheritance is an explanation for traits such as human eye color and skin color.

7. _____ The disease that is caused by two recessive genes also causes an intermediate case of the disease when only one copy of the recessive allele is present.

8. _____ One gene has several seemingly unrelated effects.

9. _____ Both of a pair of contrasting alleles that specify different phenotypes are expressed in a heterozygous individual.

10. _____ One hundred percent of persons who are homozygous recessive for the cystic fibrosis gene have cystic fibrosis disease.

11. _____ The first step to identifying genes that control common reactions to drugs is being done now.

12. ____ Some persons with camptodactyly have immobile, bent fingers on both hands. In others, the trait shows up on one hand only.

13. ____ $I^A I^B$ express both blood types

___

SELF-QUIZ

____ 1. The best statement of Mendel's principle of independent assortment is that _____. [p.380]

    a. one allele is always dominant to another
    b. hereditary units from the male and female parents are blended in the offspring
    c. the two hereditary units that influence a certain trait separate during gamete formation
    d. each hereditary unit is inherited separately from other hereditary units

____ 2. One of two or more alternative forms of a gene for a single trait is a(n) _____. [p.376]

    a. chiasma
    b. allele
    c. autosome
    d. locus

____ 3. In the grandchild generation of a monohybrid cross involving complete dominance, the expected phenotypic ratio is _____. [p.379]

    a. 3:1
    b. 1:1:1:1
    c. 1:2:1
    d. 1:1

____ 4. In a testcross, individuals with a dominant phenotype are crossed to an individual known to be _____ for the trait. [p.379]

    a. heterozygous
    b. homozygous dominant
    c. homozygous
    d. homozygous recessive

____ 5. A man with type A blood who marries a woman with type AB blood could be the father of any of the following except _____. [p.383]

    a. a child with type A blood
    b. a child with type B blood
    c. a child with type O blood
    d. a child with type AB blood

____ 6. A single gene that affects several seemingly unrelated aspects of an individual's phenotype is said to be _____. [p.382]

    a. pleiotropic
    b. epistatic
    c. mosaic
    d. continuous

____ 7. Suppose two individuals, each heterozygous for the same characteristic, are crossed. The characteristic involves complete dominance. The expected genotypic ratio of their progeny is _____. [p.378]

    a. 1:2:1
    b. 1:1
    c. 100 percent of one genotype
    d. 3:1

____ 8. If a homozygous dominant individual is mated to a homozygous recessive individual, the predicted genotypic ratio of the offspring will be _____. [p.379]

    a. 1:1
    b. 1:2:1
    c. 100 percent of one genotype
    d. 3:1

____ 9. The trait of skin color in humans exhibits _____. [p.385]

    a. pleiotropy
    b. epistasis
    c. mosaic
    d. continuous variation

CHAPTER OBJECTIVES / REVIEW QUESTIONS

1. Discuss a monohybrid cross and what type of information you can obtain from this type of experiment. [p.377]

2. _____ are units of information about specific traits; they are passed from parents to offspring. [p.376]

3. Define *allele*; how many alleles are present in the genotypes *Tt*, *tt*, and *TT*? [p. 376]

4. Having two of the same allele for a trait is a _____ condition; if the two alleles are different, this is a _____ condition. [p.376]

5. Distinguish a dominant allele from a recessive allele. [p.376]

6. _____ refers to the genes present in an individual; _____ refers to an individual's observable traits. [p.376]

7. In a _____ cross, the inheritance of a single trait is followed. [p.377]

8. Mendel's theory of _____ states that during meiosis, the two genes of each pair separate from each other and end up in different gametes. [p.377]

9. Explain why probability is useful to genetics. [p.377]

10. Define the *testcross* and cite an example. [p.378]

11. The theory of _____ states that alleles for different traits are sorted into gametes independently of each other. [p.380]

12. Explain how the steps used to predict offspring of a dihybrid cross are different from those used for a monohybrid cross. [p.381]

13. Explain *codominance*. [p.383]

14. Sickle-cell anemia is an example of _____. [pp.382-383]

15. Explain why sickle-cell trait and type AB blood are good examples of codominance. [pp.382-383]

16. Why is polydactyly said to be an incompletely penetrant genotype? [p.384]

17. Camptodactyly is said to show _____ expressivity. [p.384]

18. State the criterion that qualifies a trait to be polygenic. [pp.384-385]

INTEGRATING AND APPLYING KEY CONCEPTS

1) Solve the following genetics problem: In humans, hypotrichosis (sparse body hair) is caused by gene *h* that is recessive to normal, *H*. Pituitary dwarfism is caused by a deficiency of a growth hormone controlled by an recessive allele, *p*; normal growth is caused by dominant gene *P*. A hypotrichotic male who is homozygous for normal stature marries a woman who is homozygous for normal body hair but is a pituitary dwarf. Predict the appearance of their children. If a mating were to occur between two genotypes like those of their children, what prediction would you make concerning the appearance and the proportions of the characteristics those children might have?

2) Draw Punnett squares for each of the possible crosses between an individual who has type A blood and an individual who has type B blood. For each one state the genotype and phenotype ratios.

20

CHROMOSOMES AND HUMAN GENETICS

INTRODUCTION

Human genetic problems arise in two manners. Some are due to an incorrect number of chromosomes or chromosomes with extra or missing pieces. This chapter will describe a technique called karyotyping which is used to diagnose chromosomal abnormalities. Mistakes in control mechanisms of homologous chromosomes, in addition to genetic rearrangements that normally are associated with meiosis, contribute to some genetic disorders. This chapter will also explain the inheritance of disorders due to mutations within genes that may be passed from one generation of a family to the next. As you study these genetic problems, be aware that their presence in an individual is governed by the laws of probability, not unhealthy parental behaviors.

FOCAL POINTS

- Figure 20.3 [p.391] shows a human karyotype. As you examine chromosomal disorders in this chapter, refer to this diagram to understand how they are diagnosed.
- Figures 20.6 [p.394] and 20.11 [p.398] are examples of pedigrees. Just as karyotypes are valuable in diagnosing chromosomal disorders, pedigrees are a starting point for predicting the inheritance of gene mutations in family lines.

INTERACTIVE EXERCISES

CHAPTER INTRODUCTION (P. 389)

20.1. A REVIEW OF GENES AND CHROMOSOMES [P.390]

20.2. PICTURING CHROMOSOMES WITH KARYOTYPES [P.391]

Boldfaced Terms
karyotype _____

_____

Matching

Choose the most appropriate description for each term. [pp. 390-391]

1. _____ crossing over
2. _____ XX
3. _____ karyotype
4. _____ alleles
5. _____ genes
6. _____ meiosis
7. _____ autosomes
8. _____ linkage
9. _____ XY
10. _____ independent assortment
11. _____ diploid (2n)
12. _____ homologous
13. _____ sex chromosomes
14. _____ locus

A. X and Y chromosomes
B. Display of photographed chromosomes from a single body cell
C. Inherited units of information located on chromosomes
D. Male sex chromosomes
E. Female sex chromosomes
F. Different forms of a gene
G. Process that separates homologous chromosomes in the production of gametes
H. Location of a particular gene on a chromosome
I. Chromosomes other than the sex chromosomes
J. Exchange of segments between homologous chromosomes
K. Refers to genes physically located on the same chromosome
L. Cell with a pair of each chromosome
M. Principle stating that genes inherited from parents are sorted into gametes in new and different combinations
N. Refers to a pair of chromosomes alike in length, shape, and genes

Sequence

In the blanks beside the numbers, write the letters of the steps taken to prepare a karyotype in the correct order. [p.391]

15. _____
16. _____
17. _____
18. _____
19. _____
20. _____
21. _____

A. Colchicine is added to arrest mitosis at metaphase.
B. The cells are placed on a slide, fixed and stained.
C. Cell culture is centrifuged.
D. Cells are transferred to a saline solution causing them to swell.
E. The chromosomes are cut out and arranged by size with the sex chromosomes last.
F. Cells are cultured in a growth medium.
G. The chromosomes are photographed through a microscope.

Multiple Choice [pp. 390-391]

___ 22. The chemical colchicine is useful for karyotyping because:

    A. it forces cells to enter mitosis.
    B. it causes chromosomes to condense.
    C. it blocks cystoskeletal functions which arrests chromosomes at the metaphase plate.
    D. none of the above is correct.

___ 23. In humans, all the chromosomes can have a homologous partner in some individuals except:

    A. chromosome 21.
    B. chromosome 14.
    C. the Y chromosome.
    D. the X chromosome.
    E. none of these have homologous partners.

___ 24. A cell with a diploid number of chromosomes _____?

    A. has a pair of each chromosome
    B. is the result of meiosis
    C. means that it contains both an X and Y chromosome in all individuals
    D. always has different forms of the genes
    E. none of these are correct

___ 25. The adjacent figure represents chromosome 7 and the mutated gene is responsible for which of the following diseases.

    A. Cystic Fibrosis
    B. polydactyly
    C. Huntington's disease
    D. Trisomy
    E. none of these

20.3. THE SEX CHROMOSOMES [PP.392-393]

20.4. HUMAN GENETIC ANALYSIS [PP.394-395]

Boldfaced Terms

Human chromosome 7

X chromosome _____

_____

Y chromosome _____

_____

X-linked genes _____

_____

Y-linked genes _____

310   Chapter Twenty

X inactivation _____

_____

Barr body _____

_____

pedigree _____

_____

Huntington's disease _____

_____

genetic abnormality _____

_____

genetic disorder _____

_____

syndrome _____

_____

Complete the Table

1. Complete the Punnett square table, which brings Y-bearing and X-bearing sperm together randomly in fertilization. [p.392]

Dichotomous Choice

Answer the following questions related to the completed Punnett square. [p.392]

2. Human males transmit their Y chromosome only to their (sons / daughters).

3. Human males receive their X chromosome only from their (mothers / fathers).

4. Human mothers and fathers each provide an X chromosome for their (sons / daughters).

5. While the Y chromosome is responsible for masculinizing an embryo, most of the genes on the X chromosome deal with (sexual / nonsexual) traits.

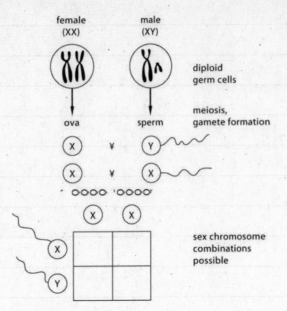

female
(XX)

male
(XY)

diploid
germ cells

meiosis,
gamete formation

ova

sperm

X   ¥   Y

X   ¥   X

X   X

X

Y

sex chromosome
combinations
possible

Matching

Choose the most appropriate description for each term.. [pp. 393-395]

6. ____ X inactivation

7. ____ Barr body

8. ____ genetic predisposition

9. ____ pedigree

10. ____ carrier

11. ____ abnormality

12. ____ genetic disorder

13. ____ syndrome

A. Designates a person who is heterozygous for a recessive trait that causes a genetic disease

B. An inherited condition causing mild to severe medical problems

C. Increased susceptibility to a disease due to inherited genes

D. Early development mechanism that randomly switches off all or most of the genes on one of a female's X chromosomes

E. Set of symptoms that usually occur together and characterize a disorder

F. A chart that is constructed to show the genetic connections among individuals

G. Condensed, inactivated X chromosome that may be seen under a light microscope

H. Deviation from the average, such as having six toes on each foot rather than five

Short Answer

14. Some women have a disorder called incontinentia pigmenti in which some patches of skin show up as lighter and others as darker. The condition is controlled by X-linked genes. This disorder never occurs in males. Explain the basis for the disorder. [p.393] _____

_____

_____

_____

_____

20.5. INHERITANCE OF GENES ON AUTOSOMES [PP.396-397]

20.6. INHERITANCE OF GENES ON THE X CHROMOSOME [PP.398-399]

20.7. PERSONALIZED MEDICINE [P.400]

Boldfaced Terms

cystic fibrosis _____

_____

autosomal recessive _____

_____

phenylketonuria (PKU) _____

_____

Tay-Sachs disease_____

_____

autosomal dominant _____

_____

Marfan syndrome _____

_____

achondroplasia _____

_____

familial hypercholesterolemia _____

_____

hemophilia _____

_____

Duchenne muscular dystrophy (DMD) _____

_____

red-green colorblindness _____

_____

amelogenesis imperfecta _____

_____

testicular feminizing syndrome _____

_____

Choice

For questions 1–15, choose from the following patterns of inheritance; some items may require more than one letter. [pp. 396-399]

a. autosomal recessive   b. autosomal dominant   c. X-linked recessive   d. X-linked dominant

1. _____ The trait is expressed in heterozygotes (of either sex).

2. _____ A genetic disease cannot be detected in a heterozygote (of either sex).

3. _____ The recessive phenotype shows up far more often in males than in females.

4. _____ Both parents of a person with a genetic disease may be heterozygous normal.

5. _____ If one parent is heterozygous for a genetic disease and the other is homozygous recessive, there is a 50 percent chance that any child of theirs will inherit the disease.

6. _____ Heterozygous normal parents can expect that one-fourth of their children will be affected by a disorder inherited in this manner.

7. _____ A son cannot inherit the recessive allele from his father, but a daughter can.

8. _____ Females can mask this gene, males cannot.

9. _____ Individuals displaying this type of disorder will always be homozygous for the trait.

10. _____ The allele behaves as a dominant in one sex and a recessive in the opposite sex.

11. _____ Heterozygous females will transmit the recessive allele to half their sons and half their daughters. Males can only transmit such traits to their daughters.

12. _____ Heterozygous females will display the trait. Males are not homozygous or heterozygous for the trait.

13. _____ All of the daughters of a man with a genetic disease will have the disease, but he cannot pass the allele to any of his sons.

Problems

14. The autosomal allele that causes albinism, *a*, is recessive to the allele for normal pigmentation, *A*. A normally pigmented woman whose father is an albino marries an albino man whose parents are normal. They have three children, two normal and one albino. Give the genotypes for each person listed. [p.396] _____

_____

_____

15. Huntington's disease is a rare form of autosomal dominant inheritance, *H*; the normal gene is *h*. The disease causes progressive degeneration of the nervous system with onset exhibited near middle age. An apparently normal man in his early twenties learns that his father has recently been diagnosed as having Huntington disorder. What are the chances that the son will develop Huntington disorder? [pp.396-397] _____

_____

_____

16. A color-blind man and a woman with normal vision whose father was color blind have a son. Color blindness, in this case, is caused by an X-linked recessive gene. If only the male offspring are considered, what is the probability that this couple's son is color blind? [pp.398-399] _____

_____

_____

17. Hemophilia A is caused by an X-linked recessive gene. A woman who is seemingly normal but whose father was a hemophiliac marries a normal man. What proportion of their sons will have hemophilia? What proportion of their daughters will have hemophilia? What proportion of their daughters will be carriers? [pp.398-399]

_____

_____

18. Explain why an individual with an XY genotype might develop as a female due to testicular feminizing syndrome. ____

_____

_____

19. What is the mechanism by which muscles degenerate in cases of Duchenne muscular dystrophy?, [p.399] _____

_____

_____

Complete the Table

20. Complete the following table by indicating whether the genetic disorder listed is caused by inheritance that is autosomal recessive, autosomal dominant, X-linked recessive, X-linked dominant, or sex influenced. [pp. 396-399]

Genetic Disorder	*Inheritance Pattern*
a. Achondroplasia	
b. Hemophilia	
c. Huntington's disease	
d. Duchenne muscular dystrophy	
e. Amelogenesis imperfecta	
f. Red-green color blindness	
g. Phenylketonuria	
h. Tay-Sachs disease	
i. Familial hypercholesterolemia	
j. Marfan syndrome	
k. Cystic fibrosis	

20.8. CHANGES IN A CHROMOSOME OR ITS GENES [PP.400-401]

20.9. CHANGES IN CHROMOSOME NUMBER [PP.402-403]

Boldfaced Terms

deletion _____

_____

cri-du-chat _____

_____

aniridia _____

_____

translocation _____

_____

duplications _____

_____

nondisjunction _____

_____

Down syndrome _____

_____

Turner syndrome_____

_____

Klinefelter syndrome _____

_____

Labeling-Matching

On rare occasions, chromosome structure becomes abnormally rearranged. Such changes may have profound effects on the phenotype of an organism. Label the following diagrams of abnormal chromosome structure as a deletion, a duplication, an inversion, or a translocation. Complete the exercise by matching and entering the letter of the proper description in the parentheses following each label. [pp.400-401]

A. The loss of a chromosome segment; an example is cri-du-chat disorder.

B. A gene sequence repeated several to many times on a chromosome.

C. The transfer of part of one chromosome to a nonhomologous chromosome; an example is when chromosome 14 ends up with a segment of chromosome 8.

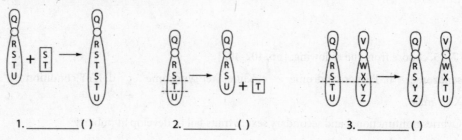

1. _____ ( )   2. _____ ( )   3. _____ ( )

Short Answer

4. Define the term gene mutation. Then list the three possible processes that may lead to a mutation and conditions associated with these processes. [pp.400-401] _____

_____

_____

_____

_____

Complete the Table

5. Complete the following table of important terms associated with chromosome number change in organisms. Choose from *monosomy, nondisjunction, aneuploidy, trisomy,* and *polyploidy.* [p.402]

Category of Change	Description
a.	New individuals do not have an exact multiple of the normal haploid set of 23 chromosomes
b.	New individuals have three or more of each type of chromosome; a lethal condition in humans
c.	One or more pair of chromosomes fail to separate during mitosis or meiosis
d.	A gamete with an extra chromosome ($n + 1$) unites with a normal gamete at fertilization; the new individual will have three of one type of chromosome ($2n + 1$)
e.	If a gamete is missing a chromosome, then the new individual resulting from fertilization will have only one of one type of chromosome ($2n - 1$)

Short Answer

6. Explain the connection between nondisjunction and aneuploidy. [p.402] _____

_____

_____

_____

Choice

For questions 7–15, choose from the following: [pp. 402-403]

a. Down syndrome   b. Turner syndrome   c. Klinefelter syndrome   d. XYY condition

7. _____ XXY male

8. _____ Ovaries nonfunctional and secondary sexual traits fail to develop at puberty

9. _____ Testes smaller than normal, sparse body hair, and some breast enlargement

10. _____ Could only be caused by a nondisjunction during sperm production by male parent

11. _____ Moderate to severe mental retardation; about 40 percent of those affected develop heart defects

12. _____ XO female; often miscarried early in pregnancy; abnormal female phenotype

13. _____ Males that tend to be taller than average but most are phenotypically normal

14. _____ Injections of testosterone reverse feminized traits but not the mental retardation or sterility

15. _____ Trisomy 21; abnormal skeleton development and muscles weaker than normal

SELF-QUIZ

_____ 1. The evidence that human females have only one functional X chromosome is provided by _____. [p.393]

    a. the Y chromosome
    b. linkage
    c. sex determination
    d. the Barr body

_____ 2. Chromosomes other than those involved in sex determination are known as _____. [p.391]

    a. nucleosomes
    b. heterosomes
    c. alleles
    d. autosomes

_____ 3. The farther apart two genes are on a chromosome, _____. [p.390]

    a. the less likely that crossing over and recombination will occur between them
    b. the greater will be the frequency of crossing over and recombination between them
    c. the more likely they are to be in two different linkage groups
    d. the more likely they are to be deleted or duplicated

_____ 4. Karyotype analysis is _____. [p.391]

    a. a means of detecting and reducing mutagenic agents
    b. a surgical technique that separates chromosomes that have failed to segregate properly during meiosis II
    c. used to detect chromosomal mutations
    d. a process that substitutes defective alleles with normal ones

_____ 5. Which of the following is not the result of an autosomal dominant gene? [p.396]

    a. Huntington's disease
    b. Cystic fibrosis
    c. Achondroplasia
    d. Familial hypercholesterolemia

_____ 6. Red/green color blindness is an X-linked recessive trait in humans. A color-blind woman and a man with normal vision have a son. What are the chances that the son is color blind? If the parents ever have a daughter, what is the probability that the daughter will be color blind? (Consider only female offspring.) [pp.398-399]

    a. 100 percent; 0 percent
    b. 50 percent; 0 percent
    c. 100 percent; 100 percent
    d. 50 percent; 100 percent

_____ 7. Suppose that a hemophilic male (X-linked recessive allele) and a female carrier for the hemophilic trait have a non-hemophilic daughter with Turner syndrome. Nondisjunction could have occurred in _____. [pp.398, 402-403]

    a. both parents
    b. neither parent
    c. the father only
    d. the mother only

_____ 8. Nondisjunction involving the X chromosome occurs during oogenesis and produces two kinds of eggs, XX and O (no X chromosome). If normal Y sperm fertilize the two types, which genotypes are possible? [pp.402-403]

    a. XX and XY
    b. XXY and YO
    c. XYY and XO
    d. XYY and YO

_____ 9. A person with $2n + 1$ chromosomes is _____. [p.402]

    a. monosomic
    b. aneuploid
    c. polyploid
    d. tetraploid

Chromosomes and Human Genetics   **319**

___ 10. Through a nondisjunction, a female has a
juvenile phenotype, 45 chromosomes,
nonfunctional ovaries, and a webbed neck.
This fits the description of _____ .
[pp.402-403]

    a. Klinefelter syndrome
    b. XYY condition
    c. Turner syndrome
    d. XXX condition

___ 11. Through nondisjunction as shown in the
figure above which of the following stages is
the point at which the "mistake" happened.
[pp.402]

    a. metaphase I
    b. anaphase I
    c. telophase I
    d. anaphase II
    e. none of these

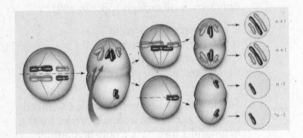

CHAPTER OBJECTIVES / REVIEW QUESTIONS

1. The units of information about heritable traits are known as _____. [p.390]

2. Diploid (2*n*) cells have pairs of _____ chromosomes. [p.390]

3. _____ are different forms of the same gene that arise through mutation. [p.390]

4. The term _____ refers to genes being located on the same chromosome. [p.390]

5. State the circumstances required for crossing over, and describe the significance of the results. [p.390]

6. Name and describe the sex chromosomes in human males and females. [p.392]

7. Distinguish between the two sex chromosomes regarding size and number of genes. [p.392]

8. Define *karyotype*; briefly describe its preparation and value. [p.391]

9. Distinguish between an X-linked gene and a Y-linked gene. [p.392]

10. Where is the gene for male sex determination located and what is it called? [p.392]

11. How are the sex chromosomes involved in sex determination? [p.392]

12. Explain the process of X inactivation and the Barr body. [p.393]

13. A(n) _____ chart or diagram is used to study genetic connections between individuals. [p.394]

14. Define the term carrier in relation to an inherited genetic disorder. [p.394]

15. In respect to genetic traits, _____ simply means deviation from the average, and a genetic _____ is an inherited condition that causes mild to severe medical problems. [p.394]

16. Explain the inheritance of an autosomal recessive trait, and give an example. [p.396]

17. Explain the inheritance of an autosomal dominant trait, and give an example. [pp.396-397]

18. Explain the inheritance of an X-linked recessive trait, and give an example. [pp.398-399]

19. Explain the inheritance of an X-linked dominat trait, and give an example. [p.399]

20. A(n) _____ trait appears more frequently in one sex than the other; pattern baldness is an example. [p.393]

21. When gametes or cells of an affected individual end up with one extra or one less than the correct number of chromosomes (not an exact multiple of the normal haploid set), it is known as _____; relate this to monosomy and trisomy. [p.402]

22. Having three or more sets of chromosomes is called _____. [p.402]

23. _____ is the failure of the chromosome pairs to separate during either mitosis or meiosis (most significant during gamete formation). [p.402]

24. Trisomy 21 is known as _____ syndrome; Turner syndrome has the chromosome constitution, _____; most _____females develop normally; XXY chromosome constitution is _____ syndrome; taller-than-average males with normal male phenotypes have the _____ condition. [pp.402-403]

25. Describe the cause and characteristics of Turner syndrome, XXX condition, Klinefelter syndrome, and the XYY condition. [pp.402-403]

INTEGRATING AND APPLYING KEY CONCEPTS

1. The parents of a young boy bring him to their doctor. They explain that the boy does not seem to be going through the same vocal developmental stages as his older brother. The doctor orders a common cytogenetics test to be done, and it reveals that the young boy's cells contain two X chromosomes and one Y chromosome. Describe the test that the doctor ordered, and explain how and when such a genetic result, XXY, most logically occurred. What treatment would the doctor most likely prescribe?

2. Solve the following genetics problem. Show rationale, genotypes, and phenotypes. A husband sues his wife for divorce, arguing that she has been unfaithful. His wife gave birth to a girl with a fissure in the iris of her eye, an X-linked recessive trait. Both parents have normal eye structure. Can the genetic facts be used to argue for the husband's suit? Explain your answer.

3. Pharmacogenetics uses the reaction of genes to specific chemicals to determine efficacy and safety of various drugs. If 2 genes are involved in the reaction of a person to a blood pressure drug describe how a doctor could design a test to determine the best reaction of 5 different drugs.

21

DNA, GENES, AND BIOTECHNOLOGY

INTRODUCTION

This chapter begins with a description of DNA, the molecule of heredity. In order to appreciate how DNA functions, an understanding of its structure is an essential starting point. This chapter examines first the structure and then the function of DNA. It then extends these concepts to some of the applications of DNA science in the area of biotechnology, an exciting although sometimes ethically controversial field.

FOCAL POINTS

- Figure 21.2 [p.409] diagrams the structure of DNA. Knowing this structure is essential to understanding genetics at any level. Labeling a diagram such as this should become as easy as writing the alphabet. With it should come the understanding that the four bases of DNA is the genetic alphabet.
- Figure 21.8 [p.410-411] and 21.12 [p.414-415] summarizes the events of DNA expression that you will be studying in most of this chapter. You should be able to describe each part of this figure as you explain the steps of making a protein based on the information within a gene.

CHAPTER INTRODUCTION [P. 407]

21.1. DNA: A DOUBLE HELIX [PP.408-409]

21.2. PASSING ON GENETIC INSTRUCTIONS [PP.410-411]

Boldfaced Terms

adenine_____

_____

guanine _____

_____

thymine _____

_____

cytosine _____

_____

base pairs _____

_____

gene _____

_____

nucleotide sequence _____

_____

DNA replication _____

_____

semiconservative replication _____

_____

DNA polymerases _____

_____

DNA repair _____

_____

xeroderma pigmentosum _____

_____

gene mutation _____

_____

base-pair substitution _____

_____

fragile X syndrome _____

_____

Short Answer

1. What benefits have been realized by genetically modifying soybean and corn crops? How might this research benefit other types of crops? [p.407] _____

_____

_____

_____

2. List the three parts of a nucleotide. In what nucleotide part do the four types of DNA nucleotides differ? [p.408]

_____

_____

_____

_____

Complete the table

3. Complete the table below showing important characteristics of the four nucleotides found in DNA. [p.408]

Name of nucleotide	Abbreviation	Single- or double-ring	Base pairs with
a.	A	e.	i.
b.	G	f.	j.
c.	T	g.	k.
d.	C	h.	l.

True/False

If false, explain why by changing one or more of the underlined words. [pp. 408-409]

4. _____ DNA is composed of <u>two</u> different types of nucleotides.

5. _____ In a nucleotide, the phosphate group is attached to the <u>nitrogen-containing base</u>.

6. _____ <u>Watson and Crick</u> are given credit for determining the structure of DNA.

7. _____ The shape of the DNA molecule is often described as a double <u>ladder</u>.

8. _____ Nucleotides are linked together by <u>covalent</u> bonds to form strands of DNA.

9. _____ The two strands of DNA are held together by <u>covalent</u> bonds between pairs of bases.

10. _____ The two DNA strands run in <u>opposite</u> directions.

11. _____ In the DNA of every species, the amount of <u>adenine</u> present always equals the amount of thymine, and the amount of <u>cytosine</u> always equals the amount of guanine (A = T and C = G).

Labeling

Identify each indicated part of the following DNA illustration. [pp.408-409]

12. _____ (deoxyribose)

13. _____

14. _____

15. _____

16. _____

17. _____

18. _____

19. _____ bond

20. _____ bond

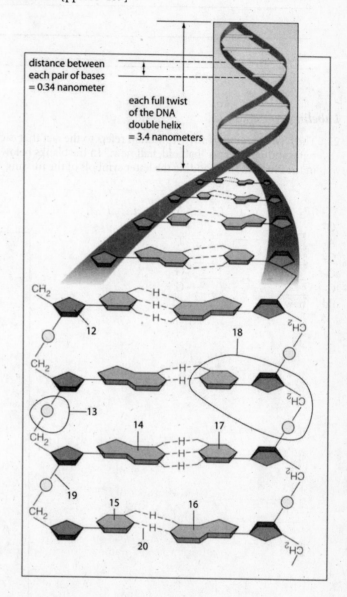

distance between
each pair of bases
= 0.34 nanometer

each full twist
of the DNA
double helix
= 3.4 nanometers

Short Answer

21. Define the word "gene" in two ways. [p.409] _____

_____

_____

_____

Labeling

22. The term *semiconservative replication* refers to the fact that each new DNA molecule resulting from the replication process is "half old, half new." In the blanks below, complete the replication of the separated strands in the illustration by adding the letter symbols of the missing nucleotide bases. [p.410]

T- ___   ___ -A
G- ___   ___ -C
A- ___   ___ -T
C- ___   ___ -G
C- ___   ___ -G
C- ___   ___ -G
old new   new old

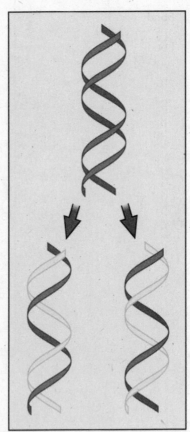

Fill-in-the-Blanks [pp. 410-411]

DNA replication occurs as the double helix unwinds and hydrogen bonds between strands are broken, exposing stretches of its nucleotide (23) _____. Each parent strand remains intact as a new companion strand is assembled on it, one (24) _____ at a time. DNA replication is said to be (25) _____ because one strand of each completed DNA molecule is "new" and the other is from the starting molecule.

Individual nucleotides are assembled into the new portion of the double helix by enzymes called (26) _____. These along with other enzymes also function in DNA (27) _____. It has been estimated that a human cell must repair breaks in a single strand of DNA up to (28) _____] times every hour. When errors that occur during DNA replication are not detected and corrected, the result is a(n) (29) _____.

DNA is also vulnerable to damage from chemicals, ionizing radiation, and (30) _____. UV light causes two neighboring (31) _____ bases to become linked. This damage can lead to the genetic disorder (32) _____ in which radiation damage to DNA cannot be fixed, placing individuals with this disease at a high risk for skin cancer.

Small-scale changes in the nucleotide sequence are called (33) _____. In a (34) _____, one base is wrongly paired with another. Sickle-cell anemia is caused by this type of mutation. A mutation in which a base has been lost is called a (35) _____. When a particular nucleotide sequence is repeated over and over, this is called a(n) (36) _____ mutation, such as causes fragile X syndrome. Regardless of the cause of a mutation, some will result in (37) _____]. They cause mutations, such as the one responsible for alkaptonuria.

While mutations can occur in any cell, they are only inherited when they take place in (38) _____ cells that give rise to gametes. Mutations may be harmful, neutral, or (39) _____ depending on how the resulting protein affects body functions.

_____ 40. Which of the following is not a nucleotide base?

    A. cytosine

    B. guanine

    C. thymine

    D. garagine

    E. none of these are nucleotides

_____ 41. Which of the following is TRUE of DNA in the cell?

    A. It is a coiled single strand of polymers

    B. It is a double strand that runs in opposite directions

    C. It is made up of amino acid subunits

    D. It uses covalent not hydrogen bonds to link two strands together.

    E. none of these are true

_____ 42. DNA has all of the following parts except:

    A. ribose sugar

    B. base

    C. phosphate group

    D. plastic grove.

    E. all of the above

_____ 43. Xeroderma pigmentosum is a genetic condition in which:

    A. skin becomes depigmented for no apparent reason

    B. high salt causes mutations

    C. ultraviolet radiation damage cannot be fixed

    D. none of these are conditions associated with this disorder.

    E. all of these are conditions associated with this disorder.

21.3. DNA INTO RNA : THE FIRST STEP IN MAKING PROTEINS [PP.412–413]

21.4. THE GENETIC CODE [P.414]

21.5. TRNA AND RRNA [PP.415]

21.6. THE THREE STAGES OF TRANSLATION [PP.416-417]

Boldfaced Terms

transcription _____

_____

ribosomal RNA (rRNA) _____

_____

messenger RNA (mRNA) _____

_____

transfer RNA (tRNA)_____

_____

RNA polymerases _____

_____

introns_____

_____

exons_____

_____

regulatory proteins _____

_____

codons_____

_____

genetic code_____

_____

start codon_____

_____

stop codons _____

_____

anticodon _____

_____

translation _____

_____

Complete the Table

1. Complete the following table to summarize the molecular differences between DNA and RNA. [p.412]

	DNA	RNA
sugar	a.	b.
bases	c.	d.

Fill-in-the-Blanks [p.412]

The two steps from genes to proteins are called (2) _____ and (3) _____ In

(4) _____, single-stranded molecules of RNA are assembled on DNA templates in the nucleus. Genes are

transcribed into (5) _____ types of RNA. Only (6) _____ is eventually translated into

protein. The other two types of RNA operate in (7) _____, the second stage of protein synthesis.

Complete the Table

8. Three types of RNA are translated from DNA in the nucleus (from genes that code only for RNA). Complete the following table, which summarizes information about these RNA molecules. [p.412]

RNA Molecule	Abbreviation	Description/Function
Ribosomal RNA	a.	
Messenger RNA	b.	
Transfer RNA	c.	

Short Answer

9. Cite three key differences between DNA replication and transcription (both processes involve DNA and occur in the nucleus). [p.412]_____

_____

_____

_____

_____

Sequence

Arrange the steps of transcription in order, with the earliest step first and the latest step last. [pp.412-413]

10. ____

11. ____

12. ____

13. ____

14. ____

15. ____

A. A termination base sequence serves as a signal to release the RNA transcript.

B. Proteins help position an RNA polymerase on the DNA so that it binds with promoter.

C. Newly formed pre-RNA is modified; introns are removed and exons are spliced together.

D. Mature mRNA leaves the nucleus.

E. A promoter is a base sequence that signals the start of a gene.

F. RNA polymerase moves along the DNA, joining RNA nucleotides together.

Completion

16. Suppose the following line represents the DNA strand that will act as a template for the production of mRNA through the process of transcription. Complete the blanks below the DNA strand with the sequence of complementary RNA bases that will represent the message carried by mRNA from DNA to a ribosome in the cytoplasm. [p.412]

TAC — AAG — ATA — ACA — TTA — TTT — CCT — ACC — GTC — ATC

_____ - _____ - _____ - _____ - _____ - _____ - _____ - _____ - _____ - _____
(transcribed single-strand of mRNA)

Translation is the second phase of protein synthesis. It occurs in the cell's cytoplasm and consists of three stages called initiation, elongation and termination. During this process, mRNA transcribed from DNA is "translated" into a linear sequence of amino acids called a polypeptide chain which is the first level of the structure of a protein. Identify each indicated part of the following illustration. Place the correct letters in the proper order for the process of translation on the blank lines that correspond to the number on the illustration. [p. 416]

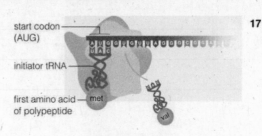

17. _____

18. _____

19. _____

20. _____

21. _____

22. _____

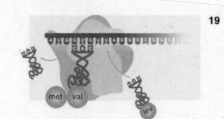

A. A peptide bond forms between the second and third amino acids.

B. A peptide bond forms between the third and fourth amino acids.

C. A peptide bonds forms between the first two amino acids.

D. The second tRNA is released and the ribosome moves to the next codon. A fourth tRNA binds the fourth codon

E. Ribosome subunits and an initiator tRNA converge on an mRNA. A second tRNA binds to the second codon.

F. The first tRNA is released and the ribosome moves to the next codon. A third tRNA binds to the third codon.

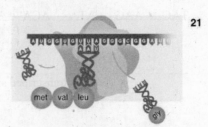

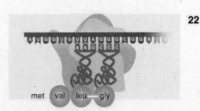

Short Answer

23. Describe the function of regulatory proteins. [p.413] _____

_____

_____

_____

Matching

Write the letter of the matching definition next to the terms below. [pp.414-415]

24. ____ codon

25. ____ sixty-four

26. ____ genetic code

27. ____ ribosome

28. ____ tRNA

29. ____ anticodon

30. ____ "stop" codons

31. ____ three-bases-at-a-time (codon)

A. Composed of two subunits made in the nucleus and sent to the cytoplasm

B. Reading frame of the nucleotide bases in mRNA

C. UAA, UAG, UGA (three of sixty-four codons)

D. A sequence of three nucleotide bases on tRNA that can pair with a specific mRNA codon

E. Contains a "hook" site that can attach to a specific amino acid

F. Name for each base triplet in mRNA

G. The number of codons in the genetic code

H. A cell's basic instructions for synthesizing proteins

Complete the Table

32. Complete the following table, which distinguishes the stages of translation. [pp.416-417]

Translation Stage	Description
a.	An initiator tRNA binds with the small ribosomal subunit and attaches to one end of the mRNA; this unit moves along the mRNA until it encounters the start codon (AUG) of the mRNA transcript; after this, a large ribosomal unit binds with a small one.
b.	A polypeptide chain forms as the mRNA strand passes between the ribosomal subunits; some proteins in the ribosome function as enzymes, joining amino acids together in the sequence dictated by mRNA codons; they catalyze the formation of a peptide bond between adjacent amino acids.
c.	A stop codon is reached and there is no corresponding anticodon; now the ribosome interacts with certain release factor proteins; this causes the ribosome as well as the polypeptide chain to detach from the mRNA; the detached chain may join the cytoplasmic pool of free proteins or enter the cytomembrane system.

33. Find your answer to question 16, and enter it on the line provided. Deduce the composition of the tRNA anticodons that would pair with the specific mRNA codons as these tRNAs deliver the amino acids that are identified here to the binding sites of the small ribosomal subunit. [pp. 415]

mRNA _____ – _____ – _____ – _____ – _____ – _____ – _____ – _____ – _____ – _____ – _____

tRNA _____ – _____ – _____ – _____ – _____ – _____ – _____ – _____ – _____ – _____

34. From the mRNA transcript in exercise 33, use Figure 21.9 in the text to identify the composition of the amino acids in the polypeptide sequence.

_____ – _____ – _____ – _____ – _____ – _____ – _____ – _____ – _____ – _____ [p.414]
(amino acids)

Matching

Match the terms with their correct definitions. [pp.412-417]

35. mRNA_____ ( )

36. tRNA_____ ( )

37. rRNA_____ ( )

38. ribosome____ ( )

39. translation___ ( )

40. transcription _(_)

A. Carries the genetic code from the nucleus to the cytoplasm where it will be translated into a protein product

B. Includes three stages: initiation, chain elongation, and chain termination

C. A nucleic acid that can transfer a specific amino acid to the mRNA/ribosome complex

D. RNA molecules are produced on DNA templates in the nucleus

E. Place where translation occurs

F. Will pick up specific amino acid for delivery to ribosome for translation

Short Answer

41. Explain what a polysome is and in what situation within a cell polysomes are valuable. [p.416] _____

_____

_____

_____

42. List two possible fates of newly formed polypeptide chains. [p.417] _____

_____

_____

_____

21.7. TOOLS FOR ENGINEERING GENES [PP.418-419]

21.8. "SEQUENCING" DNA [p.420]

Boldfaced Terms

recombinant DNA technology_____

_____

genetic engineering _____

_____

genome _____

_____

PCR (polymerase chain reaction) _____

_____

DNA sequencing _____

_____

genomics _____

_____

Fill-in-the-Blanks [pp. 418-419]

Today researchers routinely cut and splice (1) _____ from different species. Modified

molecules are then inserted into cells such as (2) _____ that can replicate genetic material and divide.

Researchers create genetic changes by using (3) _____ technology in which DNA from different (4)

_____ is cut and spliced together. These modified molecules are then inserted into bacteria or other cells

that can rapidly (5) _____ genetic material and then divide. This new technology also is the basis for (6)

_____, in which genes are altered, and inserted back into the original organism or a different one.

Bacterial (7) _____ are circular DNA molecules containing only a few genes. A(n) (8)

_____ can cut apart specific sequences of DNA. Fragments of DNA with the same (9)

"_____" ends will combine and form a recombinant DNA molecule. Foreign DNA inserted into a

plasmid is called a DNA (10) _____ because a bacterium will replicate the engineered plasmid,

producing many identical copies of the DNA. As bacteria with engineered plasmids reproduce, they (11)

_____ - make much more of - the foreign DNA.

Matching

Choose the most appropriate description for each term. [pp. 418-419]

12. _____ DNA clone
13. _____ plasmid
14. _____ "sticky" ends
15. _____ recombinant DNA
16. _____ restriction enzyme
17. _____ DNA amplification
18. _____ genome
19. _____ polymerase chain reaction

A. Splicing together DNA from different species

B. All the DNA in a haploid set of a species' chromosome

C. Rapid way to copy DNA fragments

D. Molecule that cuts a DNA strand at a specific location

E. Rapid production of DNA clones through reproduction of host cell

F. Small, circular DNA molecule that carries only a few genes

G. One of multiple, identical copies of a DNA fragment that has been inserted into a bacterial plasmid

H. DNA fragments with staggered cuts

Short Answer

20. How does PCR differ from the use of bacterial plasmids for DNA cloning? [pp.418-419]_____

_____

_____

_____

21. What is the function of primers used in PCR? [p.419]_____

_____

_____

_____

Fill-in-the-Blank [pp. 418-420]

22. _____ Determining the order of nucleotides in a gene or other piece of DNA by the use of fluorescing nucleotides

23. _____ Short stretch of radioactively labeled DNA used to find a gene of interest

24. _____ A mixture of DNA fragments containing genes of interest

21.9. MAPPING THE HUMAN GENOME [PP.420-421]

21.10. APPLICATIONS OF BIOTECHNOLOGY TO HUMAN CONCERNS [P.422-423]

21.11. ENGINEERING BACTERIA, ANIMALS, AND PLANTS [P.424]

21.12. TO CLONE OR NOT TO CLONE? [P.425]

Boldfaced Terms

Human Genome Project _____

_____

amyotrophic lateral sclerosis _____

_____

DNA chip_____

_____

gene therapy_____

_____

severe combined immune deficiency _____

_____

DNA profile _____

_____

DNA fingerprint _____

_____

therapeutic cloning _____

_____

reproductive cloning _____

_____

True/False

If false, explain why by changing one or more of the underlined words. [pp. 420-421]

1. _____ Thanks to DNA sequencing, we now know the human genome consists of <u>4.5</u> billion nucleotide bases.

2. _____ The bases of the human genome are subdivided into roughly <u>30,000</u> genes.

3. _____ The coding portions of human DNA make up about <u>80</u> percent of our DNA.

4. _____ There are over 1.4 million SNPs in the human genome, including many that result in different <u>alleles</u>.

5. _____ DNA <u>chips</u> are microarrays of thousands of DNA sequences stamped onto a glass plate, and used to pinpoint gene activity.

6. _____ The findings of the Human Genome Project may allow therapeutic drugs to be <u>customized</u> for individuals and particular situations.

7. _____ The Human Genome Project seeks to create maps of where specific genes are on chromosomes - this <u>has not</u> been successful.

8. _____ A consortium of public and private laboratories began to correlate genetic disorders with <u>individual</u> genetic differences.

9. _____ Among the concerns about biotechnology is the fear that <u>genetic profiling</u> could lead to discrimination against people seeking employment or insurance.

Fill-in-the-Blanks [pp. 422-423]

In (10) _____, one or more normal genes are inserted into a person's body to replace mutated

genes in order to correct a genetic defect. Smaller genes can be carried into a host cell by (11) _____,

while larger ones must enter in some other way. In (12) _____, DNA is integrated into exposed cells in a

laboratory culture, but it has not been very successful. (13) _____ involves inserting a gene into a virus

that then infects a host cell and carries the correct gene in. Research with SCID has used a retrovirus to insert a

338 Chapter Twenty-One

normal (14) _____ into stem cells of patients. In a trial of 11 children, (15) _____ were able to leave their isolation tents. In a similar study with cystic fibrosis, only (16) _____ percent of affected cells took up the normal gene. Gene therapy has been most successful in treating (17) _____. Genes for a(n) (18) _____ have been introduced into cultured tumor cells and the cells returned to the body. This may stimulate T cells to recognize cancerous cells and (19) _____ them. In theory, (20) _____ produced by the tumor cells may act as (21)" _____ "_ that stimulate T cells to recognize cancerous cells. At present, (22) _____ is still experimental and it will probably be years before it is used to treat and cure diseases.

Dichotomous Choice

Circle one of two possible answers given between parentheses in each statement. [pp. 422-423]

23. The unique set of (DNA fragments / coding genes) in each individual is called a DNA fingerprint.
24. The short DNA segments that differ greatly from person to person are called (repeats / riff-lips).
25. In DNA fingerprinting, scientists are looking for a (unique / common) set of DNA fragments.
26. Gene therapy aims to (create / replace) mutated genes with normal ones.

Short Answer

27. What is bioremediation and how can it help us. [p.425] _____

_____

_____

_____

_____

28. Name some benefits of transgenic organisms. [p.424] _____

_____

_____

_____

Fill-in-the-Blanks [pp. 423-425]

Any genetically engineered organism that carries foreign genes is (29) _____.

(30) _____ in bioengineered bacteria carry many human genes so that useful human

(31) _____ can be made easily. Growth hormone, (32) _____, and interferon are examples

of such products. Animal cells can be (33) _____ with foreign DNA. Using animals to produce

compounds used to treat disease eliminates the risk of transmitting blood-borne (34) _____ associated

with obtaining these products from humans. Plants may be engineered to improve traits such as

(35) _____ to a pathogen or herbicide, and to improve crop yields.

Controversy always surrounds cloning due to ethical concerns and possible outcomes after cloning takes

place. (36) _____ is no exception and it involves the use of cloned embryos to be used as a source of

embryonic (37) _____ to be used to grow replacement (38) _____ and (39) _____.

Another controversial procedure is (40) _____ which involves creating a cloned (41)

_____ that can be implanted into a woman's uterus and allowed to develop into a baby.

SELF-QUIZ

_____ 1. A DNA molecule is built from four kinds of
_____. [p.408]

  a. base pairs
  b. nitrogen-containing bases
  c. phosphates
  d. nucleotides

_____ 2. In DNA, base pairing occurs between
_____. [p.408]

  a. cytosine and uracil
  b. adenine and guanine
  c. adenine and uracil
  d. adenine and thymine

_____ 3. A single strand of DNA with the base-
pairing sequence C-G-A-T-T-G is
compatible only with the sequence
_____. [p.408]

  a. C-G-A-T-T-G
  b. G-C-T-A-A-G
  c. T-A-G-C-C-T
  d. G-C-T-A-A-C

_____ 4. During DNA replication, individual
nucleotides are assembled onto a parent
DNA strand by _____. [p.410]

  a. thymine dimers
  b. DNA polymerases
  c. ribosomes
  d. codons

_____ 5. Transcription _____. [p.412]

  a. occurs on the surface of the ribosome
  b. is the final process in the assembly of a
     protein
  c. occurs during the synthesis of RNA by
     use of a DNA template
  d. is catalyzed by DNA polymerase

_____ 6. _____ carries amino acids to
ribosomes, where they are linked into the
primary structure of a polypeptide. [p.412]

  a. mRNA
  b. tRNA
  c. An intron
  d. rRNA

7. Transfer RNA differs from other types of RNA because it _____. [p.412]

   a. transfers genetic instructions from cell nucleus to cytoplasm
   b. specifies the amino acid sequence of a particular protein
   c. carries an amino acid at one end
   d. contains codons

8. _____ is an enzyme that dominates the process of transcription. [p.412]

   a. RNA polymerase
   b. DNA polymerase
   c. Phenylketonuriase
   d. Transfer RNA

9. _____ and _____ are found in RNA but not in DNA. [p.412]

   a. Deoxyribose; thymine
   b. Deoxyribose; uracil
   c. Ribose; uracil
   d. Ribose; thymine

10. Each "word" in the DNA and mRNA language consists of _____ "letters." [p.414]

    a. three
    b. four
    c. five
    d. more than five

11. The genetic code is composed of _____ different kinds of codons. [p.414]

    a. three
    b. twenty
    c. sixteen
    d. sixty-four

12. Initiation, elongation, and termination are all stages of _____. [p.416]

    a. replication
    b. translation
    c. transcription
    d. mutagenesis

13. Small, circular molecules of DNA in bacteria ZZzare called _____. [p.418]

    a. plasmids
    b. DNA probes
    c. RFLPs
    d. cDNA

14. _____ enzymes are used to cut genes in recombinant DNA research. [p.418]

    a. Ligase
    b. Restriction
    c. Transcriptase
    d. DNA polymerase

15. Genetically engineered bacteria may become major weapons in cleaning up environmental pollution, a process called _____. [p.425]

    a. recombinant DNA technology
    b. bioremediation
    c. DNA library construction
    d. DNA sequencing

16. Amplification results in _____. [p.419]

    a. plasmid integration
    b. bacterial conjugation
    c. cloned DNA
    d. production of DNA ligase

17. A commonly used method to copy DNA is _____. [p.419]

    a. polymerase chain reaction
    b. gene expression
    c. genome mapping
    d. RFLPs

18. Transgenic species are those that carry _____. [p.424]

    a. micro-injected DNA
    b. one or more foreign genes
    c. human genes
    d. transfected cells

CHAPTER OBJECTIVES / REVIEW QUESTIONS

1. Aided by clues in X-ray images, _____ and _____ painstakingly devised a correct model of the DNA molecule. [p.408]

2. Explain what is meant by pairing of nitrogen-containing bases (base-pairing), and explain the mechanism that causes bases of one DNA strand to join with bases of the other strand. [p.408]

3. Assume that the two parent strands of DNA have been separated and that the base sequence on one parent strand is A-T-T-C-G-C; the base sequence that will complement that parent strand during replication is _____ . [p.408]

4. Explain what is meant by "Newly replicated DNA has one strand conserved from the original DNA." [p.410]

5. Briefly describe the spontaneous DNA mutations known as *base-pair substitutions, insertions, deletions*, and *expansion mutations*. [p.411]

6. Cite an example of a change in one DNA base pair that has profound effects on the human phenotype. [p.411]

7. State how RNA differs from DNA in structure and function, and indicate which features RNA has in common with DNA. [p.412]

8. _____ RNA combines with certain proteins to form a ribosome; _____ RNA carries genetic information for protein construction from the nucleus to the cytoplasm; _____ RNA picks up specific amino acids and moves them to the area of mRNA and ribosome. [p.412]

9. What mRNA code would be formed from the following DNA code: TAC-CAT-GAG-ACC-GCC-ACT? [pp.412-413]

10. Describe the process of transcription and indicate three ways in which it differs from replication. [p.412-413]

11. Transcription starts at a(n) _____ , a specific sequence of bases on one of the two DNA strands that signals the start of a gene. [p.413]

12. Actual coding portions of a newly transcribed mRNA are called _____ ; _____ are the noncoding portions. [p.413]

13. What is the function of regulatory proteins in the process of transcription? [p.413]

14. Explain the triplet nature of the genetic code. [p.414]

15. Describe the relationships between DNA and mRNA, and between mRNA and tRNA that insure the synthesis of the correct protein. [p.414]

16. Name the three stages of translation and tell what happens in each. [p.416]

17. What role do "shipping labels" play for polypeptide chains? [p.416]

18. Distinguish between recombinant DNA technology and genetic engineering. [p.418]

19. _____ are small, circular, self-replicating molecules of DNA or RNA within a bacterial cell. [p.418]

20. Some bacteria produce _____ enzymes that cut apart DNA molecules at specific base sequences; such DNA fragments may have "_____ ends" capable of base-pairing with other DNA molecules. [p.418]

21. Describe how PCR amplifies DNA segments. [p.419]

22. Explain what a DNA sequencing is, and what genomics is. [p.420]

23. State the findings of the Human Genome Project concerning the size of the human genome, as well as the percent of human DNA that codes for proteins. [p.420]

24. Define "gene therapy" and describe techniques used in gene therapy trials. [pp.422-423]

25. Why is a "DNA fingerprint" given that name? [p.423]

26. Explain what repeats are and how they are used in DNA fingerprinting. [p.423]

27. What is a transgenic organism and how are they "created" by humans? [p.424]

28. Discuss concerns that people have about biotechnology. [p.425]

29. What is the difference between therapeutic cloning and reproductive cloning? [p.425]

30. List some uses of transgenic bacteria, animals and plants. [p.424]

31. Describe each step of the process of PCR using the diagram.

A. _____

B. _____

C. _____

D. _____

E. _____

INTEGRATING AND APPLYING KEY CONCEPTS

1. Genes code for specific polypeptide sequences. Not every substance in living cells is a polypeptide. Explain how genes might be involved in the production of a storage carbohydrate (such as glycogen) that is constructed from simple sugars.

2. Discuss the ethical controversy surrounding the use of embryonic stem cells to use for research aimed at curing diseases. Briefly describe some of the diseases that this procedure is designed to cure.

22

GENES AND DISEASE: CANCER

INTRODUCTION

As you look around at your friends and family, it is sobering to think that one out of four will die of cancer as the introduction to the chapter points out. Most of us know at least one person who has dealt with and possibly died from cancer. This chapter examines cancer at both a genetic and cellular level, as well as surveying various types of cancer, their prevention and treatment. The more knowledge you have about this disease, the better prepared you will be for dealing with cancer in your own life or those of people for whom you care.

FOCAL POINTS

- Figure 22.5 [p.432] diagrams the steps in carcinogenesis at the level of a cell. All cancer starts with a single cell, but more than one mutation is required for the cell to become cancerous. This diagram will help as you explore the cellular changes that lead to cancer.
- Figure 22.10 [p.435] provides recent data on incidence of and deaths from common cancer. As you read through the information on these cancers, consider environmental and health factors that contribute to their occurrence.

INTERACTIVE EXERCISES

CHAPTER INTRODUCTION [P. 429]

22.1. THE CHARACTERISTICS OF CANCER [PP.430-431]

22.2. CANCER, A GENETIC DISEASE [PP.432-433]

22.3. CANCER RISK FROM ENVIRONMENTAL CHEMICALS [P.434]

Boldfaced Terms

hyperplasia_____

_____

tumor_____

_____

dysplasia_____

_____

cancer _____

_____

metastasis _____

_____

carcinogenesis _____

_____

proto-oncogenes _____

_____

oncogene _____

_____

tumor suppressor gene _____

_____

retinoblastoma _____

_____

inherited susceptibility to cancer _____

_____

viruses _____

_____

chemical carcinogens _____

_____

radiation _____

_____

breakdowns in immunity _____

_____

Matching

Choose the most appropriate description for each term. [pp. 430-431]

1. ____ dysplasia
2. ____ neoplasm
3. ____ malignant
4. ____ cancer cell
5. ____ hyperplasia
6. ____ metastasis
7. ____ angiogenin
8. ____ benign

A. Growth factor secreted by cancer cells; promotes blood vessel growth around cancer cells

B. Cell lacking clear structural specializations; displays abnormally large nuclei, less cytoplasm, and disorganized cytoskeleton

C. Cancer cells breaking away from a tumor and establishing new cancer sites

D. Refers to a noncancerous, slow-growing, and well- differentiated tumor; often enclosed by a capsule

E. Overgrowth of cells leading to a tumor

F. "Bad form"; abnormal changes in size, shape and organization of cells in a tissue

G. Means "new growth"; a defined enlarged mass of tissue called a tumor

H. A cancer with the ability to metastasize and cause harm

Dichotomous Choice

Circle one of two possible answers given between parentheses in each statement. [pp. 432-433]

9. The transformation of a normal cell into a cancerous one is called (metastasis / carcinogenesis).

10. (Oncogenes / Proto-oncogenes) are normal genes regulating cell growth and development.

11. A DNA segment capable of inducing cancer in a normal cell is a(n) (oncogene / proto-oncogene).

12. An oncogene (does / does not) respond to controls over cell division.

13. An oncogene acting alone (does / does not) cause malignant cancer

14. The onset of cancer requires mutations in several genes, including mutation of at least one (tumor suppressor gene / proto-oncogene).

15. Retinoblastoma, or childhood eye cancer, is likely to develop when a child inherits (one / two) normal copies of a tumor suppressor gene.

16. Women have a high risk of developing breast cancer if they have inherited a mutated BRCA1or BRCA2 gene because these are (oncogenes / tumor suppressor genes).

17. The p53 protein stops cell (growth / division) when cells are damaged, preventing mutations from being passed on to daughter cells.

18. When p53 mutates, it cannot prevent mutated cells from dividing, and the resulting faulty protein seems to (inhibit / activate) an oncogene.

19. The mutation of an oncogene or a related suppressor gene, or a change in chromosome structure that moves an oncogene into a new position can cause the oncogene to be (expressed / inhibited).

Labeling

Identify each numbered part of the following illustration that traces the steps of carcinogenesis. [p.432]

20. _____

21. _____

22. _____

23. _____

24. _____

Matching

Match the correct type of cancer to the chemical/substance that causes it. Place the correct letter on the blank line preceding the substance. [p. 434]

25. _____ Benzene

26. _____ Asbestos

27. _____ Hydrocarbons

28. _____ Vinyl chloride

29. _____ Nickel

30. _____ Various solvents

A. Lung, epithelial linings of body cavities

B. Lung, nasal epithelium

C. Leukemias

D. Skin, lung

E. Liver, various connective tissues

F. Bladder, nasal epithelium

Choice

Choose one of these routes to cancer for each of the statements below. [pp. 432-433]

a. Inherited susceptibility   b. Viruses   c. Chemical carcinogens   d. Radiation   e. Immunity breakdowns

31. _____ Sources include tanning lamps, diagnostic x-rays, radioactive materials

32. _____ May switch on a proto-oncogene by inserting into host cell DNA

33. _____ Plays a major role in about 5% of cancers, including breast, colorectal and lung cancer

34. _____ Factor in increase in cancers as we grow older

35. _____ Examples are asbestos, vinyl chloride, and benzene, aflotoxin produced by fungi

36. _____ May carry an oncogene into the host cell DNA

37. _____ Some are pre-carcinogens, others are promoters

38. _____ Passed in gamete to next generation

39. _____ Antigens on the surface of cancer cells do not trigger an immune response

40. _____ Sun exposure is the greatest risk in this category

Fill-in-the-Blanks [p. 434]

Factors in our environment lead to about (41) _____ of all cancers. Government statistics indicate that about (42) _____ percent of the food in American supermarkets contains detectable residues of one or more of the active ingredients used in (43) _____. Residues of pesticides can be removed from food by washing, but it is difficult to avoid contact with pesticides used in community (44) _____ programs to control mosquitoes. In addition to agricultural chemicals, (45) _____ chemicals have also been linked to cancer. Based on research using the bacterium *Salmonella* as a test organism, it appears that more than 80 percent of known cancer-causing chemicals produce (46) _____. The National Academy of Sciences has warned that the active ingredients of many fungicides, herbicides and insecticides used in the U.S. have the (47) _____ to cause cancer.

Multiple Choice

_____ 48. A tumor can be defined as:

A. A slow growing fluid filled swelling
B. Death of tissue resulting in swelling
C. A mass of tissue that can either be benign or malignant
D. An allergic reaction

_____ 49. Another term for a tumor can be:

A. dysplasia
B. hyperplasia
C. metastasis
D. oncogene

_____ 50. Cancerous cells are said to go through changes in size shape and organization. Which of the following terms is used to describe this phenomenon?

A. dysplasia
B. hyperplasia
C. metastasis
D. oncogene

_____ 51. The genetic cause of many cancers is induced when a virus or chemical affects a specific structure. Which of the following terms represents that structure?

A. dysplasia
B. hyperplasia
C. metastasis
D. oncogene

____ 52. Chemicals that are known to stimulate cancer development are called:

    A. carcinogens
    B. hyperplasia
    C. metastasis
    D. oncogene

22.4. SOME MAJOR TYPES OF CANCER [P.435]

22.5. CANCER SCREENING AND DIAGNOSIS [P.436-437]

22.6. CANCER TREATMENT AND PREVENTION [P.438-439]

Boldfaced Terms

sarcomas _____

_____

carcinomas_____

_____

lymphomas _____

_____

leukemias_____

_____

Gliomas _____

_____

tumor markers_____

_____

medical imaging_____

_____

biopsy_____

_____

chemotherapy _____

_____

adjuvant therapy _____

_____

radiation therapy _____

_____

Completion

Complete the following statements of the seven warning signs of cancer. [p.437]

1. Change in _____ or _____ habits and function

2. A _____ that does not heal

3. Unusual _____ or _____ discharge

4. Thickening or _____

5. _____ or difficulty swallowing

6. Obvious change in a _____ or _____

7. Nagging _____ or _____

Matching

Choose the most appropriate description for each term. [pp. 436-439]

8. _____ chemotherapy

9. _____ medical imaging

10. _____ monoclonal antibodies

11. _____ tumor markers

12. _____ adjuvant therapy

13. _____ DNA probe

14. _____ cancer screening

15. _____ interferon

16. _____ biopsy

A. Segment of radioactively labeled DNA used to locate gene mutations or alleles associated with some types of inherited cancers

B. The use of drugs to kill cancer cells

C. Definitive cancer detection tool; a small piece of tissue is removed from the body and analyzed

D. Substances such as HCG or PSA produced by cancer cells or normal cells in response to cancer; detected in blood tests

E. Activates cytotoxic T cells and natural killer cells that recognize and kill various types of cancer

F. Diagnosis using MRI, X rays, ultrasound, and CT

G. A treatment combining surgery and a less toxic dose of chemotherapy

H. Examples are mammograms, Pap test, testicle self-examination, and breast self-examination

I. Pinpoint location and sizes of tumors of the brain, bone, colon and some other tissues

Fill-in-the-Blanks

Cancers of connective tissues (muscle and bone) are (17) _____ [p.435]. Cancers that arise

from epithelium, including cells of the skin and epithelial linings of internal organs, are known as (18)

_____ [p.435]. (19) _____ [p.435] develop in glial cells of the brain. (20) _____

[p.435] are cancers of lymphoid tissues in organs such as lymph nodes. Cancers of blood-forming regions such as

stem cells in bone marrow are (21) _____ [p.435].

SELF-QUIZ

____ 1. A defined mass of tissue known as a tumor
may not be called a _____. [p.430]
a. benign tumor
b. neoplasm
c. malignant tumor
d. capsule

____ 2. Cancer cells that break away from a
primary tumor and migrate to establish
new cancer sites are said to be undergoing
_____. [p.431]
a. activation
b. dysplasia
c. carcinogenesis
d. metastasis

____ 3. Of the following, which one is not
characteristic of cancer cells? [p.430]
a. normal appearance
b. uncontrolled growth
c. metastasis
d. lack strong cell-to-cell adhesion

____ 4. The p53 gene _____. [p.433]
a. initiates metastasis
b. mutates and becomes a malignant cell
c. is a suppressor gene effective in many
tissue types
d. is a type of oncogene

____ 5. A DNA segment capable of inducing cancer
in a normal cell is a(n) _____. [p.433]
a. tumor suppressor gene
b. oncogene
c. p53 gene
d. proto-oncogene

____ 6. Carcinogenesis is _____. [p.432]
a. the transformation of a normal cell into
a cancerous one
b. the production of carcinogens
c. the dying process of tumor suppressor
genes
d. a breakdown in immunity that leads to
cancer

____ 7. Of the following, which may serve as a
trigger for carcinogenesis? [p.433]
a. viruses
b. chemicals
c. radiation
d. heredity
e. all the above can trigger carcinogenesis

____ 8. _____ is a cancer screening technique.
[p.436]
a. Chemotherapy
b. MRI
c. Immune therapy
d. Adjuvant therapy

____ 9. Sarcomas are cancers of _____. [p.435]

a. lymph nodes
b. connective tissues
c. blood-forming regions
d. a gland or its ducts
e. epithelium

Choice

For questions 10–14, select the correct type of cancer from the following list: [pp. 434-436]

      a. Oral and lung cancer
      b. Cancer of the skin
      c. Cancer of the breast
      d. Cancer of the blood
      e. Cancer of the prostate gland

10. ____ PSA is a test for _____. [p.436]

11. ____ A malignant melanoma is a(n) _____. [p.430]

12. ____ Low-dose mammography detects, and lumpectomy may treat _____. [p.436]

13. ____ Smoking and smokeless tobacco cause _____. [p.434]

14. ____ Leukemia is a_____. [p.435]

CHAPTER OBJECTIVES / REVIEW QUESTIONS

1. If cells overgrow, the result is a defined mass of tissue called a(n) _____. [p.430]

2. _____ is an abnormal change in the sizes, shapes, and organization of cells in a tissue. [p.430]

3. List the characteristics that distinguish benign and malignant tumors. [p.430]

4. A(n) _____ is a gene that can induce cancer in a normal cell. [p.432-433]

5. In addition to an oncogene, the onset of malignant cancer requires the mutation of at least one _____ gene. [p.433]

6. Explain the importance of the p53 gene. [p.433]

7. List various factors that may trigger expression of an oncogene. [p.433]

8. List five routes to carcinogenesis other than oncogenes. [p.433]

9. Give brief statements that relate cancer to agricultural and industrial chemicals. [p.434]

10. Define *cancer screening, tumor markers, biopsy,* and *medical imaging.* [p.436]

11. List the recommended cancer screening tests. [p.436]

12. _____ is the use of drugs to kill dividing cells. [p.438]

13. Why is loss of hair cells, stem cells, lymphocytes, and epithelial cells a side effect of chemotherapy? [p.438]

14. List six American Cancer Society strategies for limiting your risk of cancer. [p.439]

15. Identify these types of cancers by their source tissue: sarcomas, carcinomas, adenocarcinomas, lymphomas, and leukemias. [p.435]

INTEGRATING AND APPLYING KEY CONCEPTS

1) A large number of chemicals, environmental pollutants, and several types of radiation are recognized as being carcinogenic. Can you think of a common effect these various and different carcinogens might have on a cell to induce cancer?

2) Considering the genetic changes that lead to cancer, why do you think that cancer is more common in older people than in younger people?

23

PRINCIPLES OF EVOLUTION

INTRODUCTION

The theory of evolution is a major keystone in the study of biology. It explains the diversity of life on earth today and the history and origins of biological life forms. This chapter begins with the development of that theory by Charles Darwin in the 19th century, then outlines evidences, such as the fossil record, radiometric dating, biogeography, comparative morphology and biochemistry, to support the theory. The evolution and geographic distribution of the human species is discussed at length. Lastly, the chemical origins of life on earth are examined.

FOCAL POINTS

- Figure 23.1, animated, [p.444] depicts Charles Darwin and the travels of the HMS *Beagle*
- Figure 23.3a [p.447] demonstrates the concept of divergence over time and its effect on speciation
- Figure 23.4 [p.448] depicts fossilized remains
- Figure 23.6, animated, [p.449] illustrates the effects of plate tectonics on the relative position of the earth's land masses over time
- Figure 23.7, animated, [p.450] compares the homologous structures of several vertebrate forelimbs
- Figure 23.8, animated, [p.451] demonstrates the embryological similarities among diverse vertebrate species
- Figure 23.9, animated, [p.451] shows how adult traits may vary from the basic juvenile form
- Figure 23.10, animated, [p.451] illustrates vestigial body parts
- Figure 23.12 [p.453] is a geological timeline showing mass extinctions of species during the various time periods of the earth's history
- Table 23.1[p.454] shows the taxonomic hierarchy of the human species
- Figure 23.15 [p.454] illustrates the relationships of various members of the primate lineage.
- Figure 23.16, animated, [p.455] illustrates the changes in primate skeletal structure during the evolution of bipedal walking
- Figure 23.19 [p.457] illustrates the routes and times of the dispersal of the human species
- Figure 23.22 [p.459] outlines one hypothesis of the origin of biological molecules and cells

INTERACTIVE EXERCISES

CHAPTER INTRODUCTION [p.443]

23.1. A LITTLE EVOLUTIONARY HISTORY [p.444]

23.2. A KEY EVOLUTIONARY IDEA – INDIVIDUALS VARY [p.445]

Boldfaced Terms

evolution _____

_____

microevolution _____

_____

macroevolution _____

_____

population _____

_____

gene pool _____

_____

Complete the Table

Complete the following table of key figures and events in the development of the theory of evolution by natural selection. [p.444]

Key Figures / Events	Importance to Synthesis of Evolutionary Theory
1.	A 22 year old with a degree in theology who preferred being in the natural world
2.	British ship that carried Darwin on a five-year voyage (as a naturalist) around the world
3.	Islands off the western coast of South America where Darwin made observations that would lead to his ideas about evolution
4.	Clergyman and economist who proposed that any population tends to outgrow its resources, and its members must compete for what is available
5	Evolutionary process that Darwin proposed after considering his observations and the ideas of other thinkers

Matching

Choose the most appropriate description for each term. [pp.444-445]

6. _____ gene pool

7. _____ behavioral traits

8. _____ evolution

9. _____ macroevolution

10. _____ microevolution

11. _____ morphological traits

12. _____ physiological traits

13. _____ population

A. Cumulative genetic changes that may give rise to new species

B. The collective genes with their alleles present in a population

C. Functions of body structures

D. A group of individuals of the same species occupying a given area

E. Responses to certain basic stimuli

F. The large-scale patterns, trends, and rates of change among groups of species

G. General form and appearance

H. Genetic changes in lines of descent through successive generations

23.3. MICROEVOLUTION: HOW NEW SPECIES ARISE [pp.446-447]

Boldfaced Terms

natural selection _____

_____

adaptation _____

_____

genetic drift _____

_____

gene flow _____

_____

species _____

_____

speciation _____

_____

Short Answer [p.446]

1. State the six main points of Darwin's Theory of Evolution by Natural Selection. _____

_____

_____

_____

_____

_____

_____

Dichotomous Choice

Circle one of two possible answers given between parentheses in each statement. [pp.446-447]

2. A unit of one or more populations of individuals that can interbreed under natural conditions and produce fertile offspring is a(n) (ecosystem / species).

3. (Natural selection / Mutations) is/are the source of alleles that provide variation in a species.

4. (Gene flow / Natural selection) is the difference in survival and reproduction that has occurred among individuals that differ in one or more traits.

5. Charles Darwin presented his ideas in a book called (*On the Origin of Species* / *The Evolution of Man*).

6. Evolution and natural selection are considered by most scientists to be (hypotheses / fundamental principles).

7. Through evolution over time, organisms come to have characteristics that suit them to conditions in a particular environment; this is (reproductive isolation / adaptation).

8. The blonde hair and blue eyes of Finns, along with many genetic disorders most common in the Finnish population can be explained by (natural selection / the founder effect).

9. The founder effect is an example of rapid change leading to unique features in a small population, a phenomenon known as (genetic drift / gene flow).

10. Allele frequencies can change as individuals leave a population or new individuals enter; this is referred to as (gene flow / genetic drift).

11. Reproductive isolation develops when (genetic drift / gene flow) between two populations stops.

12. Two species are reproductively isolated when they have enough genetic differences that they cannot produce (mating hormones / fertile offspring).

13. Different breeding seasons, different mating rituals, and changes in body structures that prevent mating serve as (hybridization / isolating) mechanisms that reduce the likelihood of interbreeding between populations.

14. The build-up of differences in allele frequencies among isolated populations is called (convergence / divergence).

15. The emergence of new species through many small changes in form over long spans of time is called (punctuated equilibrium / gradual speciation).

16. A (neutral trait / beneficial trait) improves an individual's chances of surviving and reproducing..

Short Answer

The following illustration depicts the divergence of one species into two as time passes. Answer these three questions referring to the illustration. [p.447]

17. In what area(s) is/are there distinctly one species? _____

18. Between what letters does the divergence begin? _____

19. What letter represents the time when divergence is probably complete? _____

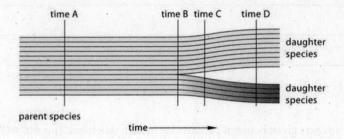

23.4. LOOKING AT FOSSILS AND BIOGEOGRAPHY [pp.448-449]

23.5. COMPARING THE FORM AND DEVELOPMENT OF BODY PARTS [pp.450-451]

23.6. COMPARING GENETICS [p.452]

Boldfaced Terms

fossil _____

_____

radiometric dating _____

_____

biogeography _____

_____

comparative morphology _____

_____

homologous structures _____

_____

analogous structures _____

_____

Matching

Choose the most appropriate answer that matches each example of evidence supporting macroevolution. [pp.448-452]

1. ____ analogous structures
2. ____ biogeography
3. ____ comparative biochemistry
4. ____ cytochrome c
5. ____ comparative morphology
6. ____ fossilization
7. ____ homologous structures
8. ____ Pangaea
9. ____ radiometric dating
10. ____ plate tectonics
11. ____ stratification
12. ____ vestigial structures

A. An electron transport protein used to compare evolutionary divergence of species

B. Body parts of close evolutionary relatives are modified in different ways as adaptations to different environments

C. Uses decay rate of radioactive isotopes to determine the age of volcanic rocks

D. Study of world distribution of plants and animals

E. Involves burial in sediments or volcanic ash, water infiltration, chemical changes, and pressure

F. Uses information contained in patterns of body form to reconstruct evolutionary history

G. Gene sequencing and amino acid sequence comparison, especially of neutral mutations

H. A massive supercontinent 240 million years ago

I. Sediments deposited in superimposed layers

J. Movement of floating segments of the earth's crust; affects dispersal of species

K. Ear-wiggling muscles in humans, pelvic girdle in pythons, and the human coccyx

L. Body parts of organisms without a common recent ancestor that resemble each other due to adaptation to a common environment

True-False

Write True or False in the blank preceding the statement. [p.448]

13. _____ We will not be able to recover the fossils for most extinct species.

14. _____ Movements of Earth's crust have wiped out parts of the fossil record.

15. _____ Jellyfish and other soft-bodied organisms are well represented in the fossil record.

16. _____ Finding fossils of small populations producing few offspring is less likely than finding fossils of large populations with a high reproductive rate.

17. _____ The environment of a species does not seem to be a factor in how well it is represented in the fossil record.

18. _____ Organisms are more likely to be preserved if they are buried in the presence of oxygen. \

23.7. HOW SPECIES COME AND GO [pp.452-453]

23.8. EVOLUTION FROM A HUMAN PERSPECTIVE [pp.454-455]

23.9. EMERGENCE OF EARLY HUMANS [pp.456-457]

Boldfaced Terms

extinction _____

_____

mass extinction _____

_____

adaptive radiation _____

_____

genus _____

_____

Precision grip and power grip _____

_____

Improved daytime vision _____

_____

Changes in dentition _____

_____

Changes in brain and behaviors _____

_____

bipedalism _____

_____

Matching

Choose the most appropriate description for each term. [pp.452-454]

1. ____ adaptive radiation
2. ____ binomial system
3. ____ extinction
4. ____ mass extinction
5. ____ background extinction

A. A long-standing means of naming species; an organism's scientific name in two parts

B. An abrupt, widespread rise in extinction rates above the background level; a global event in which major groups of organisms are simultaneously wiped out

C. The permanent loss of a species

D. New species of a lineage fill a wide range of habitats during bursts of evolutionary activity

E. A relatively steady rate of species' disappearance over time

Dichotomous Choice

Circle one of the two possible answers given between parentheses in each statement. [pp.454-457]

6. (Gorillas, Bonobos) are our closest living evolutionary relative.

7. (Primates / Hominids) evolved around 60 million years ago.

8. Through alterations in hand bones, the fingers of primates can be wrapped around objects in (opposable / prehensile) movements.

9. The thumb and tip of each finger of primates can touch in (opposable / prehensile) movements.

10. Primates evolved (forward-directed eyes / an eye on each side of the head) that allow detecting shapes and movements in three dimensions.

11. Further evolution of primate eyes allows them to respond to variations in color and light intensity, an advantage for life in the (grasslands / trees).

12. Changes in the dentition of early primates eventually led to specialization for a(an) (carnivorous / omnivorous) diet.

13. Humans evolved (bow-shaped / rectangular) jaws and (long canine teeth / smaller teeth of about the same length).

14. In many primate lineages, parents started to invest more effort in (fewer / more) offspring.

15. As parents formed stronger bonds with their young and maternal care became intense, the learning period grew (shorter / longer).

16. The linking of brain modifications and behavioral complexity is most evident in the parallel evolution of the human (hand / brain) and culture.

17. (Culture / Education) is the sum total of behavior patterns of a social group, passed between generations by learning and symbolic behavior—especially language.

18. The capacity for (emotions / language) arose among ancestral humans through changes in the skull bones and expansion of parts of the brain.

19. The habitual two-legged gait peculiar to the evolved, reorganized human skeleton is called (opposable movement / bipedalism).

20. Compared with monkeys and apes, humans have a (longer / shorter), S-shaped backbone.

21. The position and shape of the human backbone, knee and ankle joints, and pelvic girdle are the basis of (bipedalism / prehensile movement).

22. By 36 million years ago, tree-dwelling primates called (hominoids / anthropoids) had evolved in tropical forests; they included ancestors of monkeys, apes, and humans.

23. Between 23 and 5 million years ago, the first apelike forms adaptively radiated through Africa, Asia and Europe due to changes in (climate/ genetic drift).

24. By 7 million years ago, continents began to assume their current positions, climates became cooler and drier and many of our ancestral species became extinct. One survivor of this extinction led to the (great apes/ anthropoids).

25. Another survivor of this extinction gave rise to the first (hominins / anthropoids).

26. The first known specimen of *Australopithecus afarensis*, an early hominin, is nicknamed (Lucy / Linus).

27. *Homo habilis* (did / did not) use tools.

28. The first *Homo* species to leave Africa was *Homo (erectus / sapiens)*.

29. Neanderthals were present in Europe and the Near East until about (18,000 / 80,000) years ago.

30. The (multiregional / African emergence) model says that *Homo sapiens* evolved in Africa and replaced *Homo erectus* in other areas.

31. The (multiregional / African emergence) model says that *Homo sapiens* evolved from different dispersed populations of Homo erectus.

32. Since the emergence of modern humans, evolution has been almost entirely (biological / cultural) for the past 40,000 years.

Completion

Select from the following choices to complete the table of human classification. [p.454]

Animalia   Chordata   Eukarya   Hominidae   *Homo*   Mammalia   Primates   *sapiens*

Classification of Humans

Domain	33.
Kingdom	34.
Phylum	35.
Class	36.
Order	37.
Family	38.
Genus	39.
Species	40.

23.10. EARTH'S HISTORY AND THE ORIGIN OF LIFE [pp.458-459]

Boldfaced Terms

chemical evolution _____

_____

True-False

Place a "T" for True or an "F" for False in the space before the following questions. [pp.458-459]

1. _____ The first atmosphere contained gaseous oxygen.

2. _____ Liquid water was essential for the formation of cell membranes.

3. _____ Without an oxygen-rich atmosphere, the organic compounds that led to life would never have formed.

4. _____ The first living cells emerged around 2.5 billion years ago.

5. _____ The energy to drive chemical reactions that yielded organic molecules might have come from aerobic respiration.

6. _____ Complex compounds may have formed on clay layers in tidal flats.

7. _____ Complex compounds might have formed at deep sea vents.

8. _____ The first "molecule of life" was possibly not DNA but ATP.

9. _____ Membrane-bound sacs resembling cell membranes have been formed in laboratories.

10. _____ "Nanobes" may be the emergence of a new, highly evolved, life form.

SELF-QUIZ

____ 1. The term _____ refers to cumulative genetic changes that may give rise to new species. [p.445]

    a. macroevolution
    b. mutation
    c. microevolution
    d. variation

____ 2. _____ applies to the large-scale patterns, trends, and rates of change among groups of species. [p.445]

    a. Macroevolution
    b. Natural selection
    c. Microevolution
    d. Variation

____ 3. Evolution occurs when there is change in the genetic makeup of _____. [p.445]

    a. individuals
    b. gametes
    c. populations
    d. ecosystems

____ 4. Charles Darwin _____. [p.444]

    a. was a theology student
    b. was a naturalist who loved to watch wildlife
    c. travelled the world on the HMS *Beagle*
    d. all of the above

5. Which of the following statements is incorrect? According to the theory of evolution by natural selection _____. [p.446]

    a. over time, organisms adapt to their environment
    b. individuals of a population vary in the allelic forms of their genes
    c. the relative proportion of allelic forms of a gene do not change over time
    d. some allelic versions of a trait are more advantageous than others

6. When the relative numbers of different alleles in a gene pool change randomly through the generations because of chance events alone, that change is called _____. [p.446]

    a. natural selection
    b. genetic drift
    c. gene flow
    d. mutation

7. The source of new alleles for natural selection to work with is _____. [p.446]

    a. natural selection
    b. genetic drift
    c. gene flow
    d. mutation

8. In the model known as _____, new species emerge through many small changes in form over long expanses of time. [p.447]

    a. reproductive isolation
    b. punctuated equilibrium
    c. gradual speciation
    d. comparative evolution

9. Divergence may be the first stage on the road to _____, the process by which species originate. [p.447]

    a. natural selection
    b. speciation
    c. the founder effect
    d. genetic variation

10. Related species remain alike in many ways; these shared but diverged characteristics of a common genetic plan are _____. [p.450]

    a. analogous
    b. vestigial
    c. homologous
    d. adaptive

11. The physical record for the long history of life comes from _____. [p.448]

    a. comparative morphology
    b. fossils
    c. comparative embryology
    d. comparative biochemistry

12. Modern studies of _____ indicate that all present-day continents were once part of the supercontinent called Pangea. [p.449]

    a. adaptive radiation
    b. plate tectonics
    c. meteor impact
    d. mass extinction

13. Many new mammal species arose and moved into habitats vacated by dinosaurs during a(n) _____. [pp.452-453]

    a. adaptive radiation
    b. bottleneck
    c. meteor impact
    d. reverse extinction

14. When gene flow between two populations stops, it leads to _____. [p.447]

    a. adaptive radiation
    b. reproductive isolation
    c. punctuated equilibrium
    d. natural selection

15. Which of the following *is not* an adaptive characteristic of human evolution? [pp.454-455]

    a. precision and power grip
    b. enhanced nighttime vision
    c. fewer offspring and lengthened parental attention
    d. bipedal walking

___ 16. The multiregional model and the African emergence model are attempts to explain the origins of _____. [pp.456-457]

    a. *Homo sapiens*
    b. *Homo erectus*
    c. *Homo habilis*
    d. *Homo neanderthalensis*

___ 17. Which of the following is the most ancient of our genus? [p.456]

    a. *Australopithecus afarensis*
    b. *Homo erectus*
    c. *Homo sapiens*
    d. *Homo habilis*

___ 18. What two molecules that were not present on early Earth are essential for life now? [p.458]

    a. H_2 and N_2
    b. O_2 and N_2
    c. H_2O and CO_2
    d. O_2 and H_2O

CHAPTER OBJECTIVES/REVIEW QUESTIONS

1. _____ refers to genetic changes in lines of descent over time. [p.444]

2. Briefly describe Charles Darwin's early life and his contribution to evolutionary theory. [p.444]

3. Contrast the definitions of microevolution and macroevolution. [p.445]

4. List the three types of traits possessed by members of a population. [p.445]

5. Relate the term "gene pool" to the definition of a population. [p.445]

6. _____ are the source of new alleles, hence of the heritable variation in traits. [p.446]

7. Define natural selection. [p.446]

8. Define each of the following evolutionary forces: genetic drift, gene flow, reproductive isolating mechanisms. [pp.446-447]

9. The occurrence of chance events in bringing about changes in allele frequencies in small populations is known as _____. [p.446]

10. Name the unit whose individuals can interbreed under natural conditions and produce fertile offspring. [p.447]

11. Why does the explanation of life's history include both the model of evolution in bursts and the gradualism model? [p.447]

12. Briefly define and cite an example of each of the following lines of evidence supporting macroevolution: the fossil record, biogeography, comparative morphology, homologous and analogous structures, comparative embryology, vestigial structures, and comparative biochemistry. [pp.448-452]

13. A(n) _____ is an abrupt, widespread rise in extinction rates above the background extinction level. [p.452]

14. In a(n) _____, new species of a lineage fill a wide range of habitats during bursts of microevolution. [pp.452-453]

15. Name and briefly discuss the five trends in human evolution. [pp.454-455]

16. Distinguish between precision grip and power grip. [pp.454-455]

17. Define culture. [p.455]

18. Briefly discuss the species of genus Homo that exist in the fossil record, as well as the two ideas for the evolution of modern day human populations. [pp.456-457]

19. Describe the conditions of early Earth, and explain how they contributed to the origin of life. [p.458]

20. Which possible aspects of the evolution of life have been demonstrated in laboratory research? [p.459]

INTEGRATING AND APPLYING KEY CONCEPTS

1. Imagine that in the next decade three more Chernobyl-type disasters happen, the oceans acquire critical levels of carcinogenic pesticides that work their way up the food chains, and the ozone layer shrinks dramatically in the upper atmosphere. Describe the macroevolutionary events that you believe might follow.

2. As Earth becomes increasingly loaded with carbon dioxide and various industrial waste products, how do you think life forms on Earth will evolve to cope with these changes?

3. Consider the morphology, habitat and life style of the penguin, a member of the Avian class of vertebrates. What adaptations do penguins exhibit that are quite divergent from other birds? What advantage have these adaptations conferred on penguins in their environment?

24

PRINCIPLES OF ECOLOGY

INTRODUCTION

This chapter examines both the ecological interrelationships of organisms with their environment and the biogeochemical cycles on which life depend. Understanding how energy flows and nutrients cycle between the biotic and abiotic parts of the biosphere is central to the study of ecology. This knowledge is essential to comprehending the environmental issues you will encounter in chapter 25.

FOCAL POINTS

- Understanding the boldfaced terms found in sections 24.1, 24.2, and 24.3 [pp.464-468] will provide you with a working terminology of the basics of ecological study.
- Figure 24.4, animated, [p.466] diagrams the flow of energy and cycling of nutrients in ecosystems.
- Figure 24.5 [p.466] shows a food chain.
- Figure 24.6, animated, [p.467] demonstrates the interrelationships within a food web. Compare it with Figure 24.5 [p.466] of a food chain to understand the complexity of the feeding relationships within an ecosystem.
- Figure 24.8 [p.468] illustrates the relationship of trophic levels in energy and biomass pyramids
- Figure 24.9 [p.469] diagrams the hydrologic cycle
- Figure 24.10, animated, [pp.470-471] illustrates the complex carbon cycle
- Figure 24.12 [p.472] details the organisms and processes involved in the nitrogen cycle

INTERACTIVE EXERCISES

CHAPTER INTRODUCTION [p.463]

24.1. SOME BASIC PRINCIPLES OF ECOLOGY [pp.464-465]

Boldfaced Terms

ecology _____

_____

habitat _____

_____

community _____

_____

biome_____

_____

ecosystem _____

_____

niche _____

_____

succession _____

_____

Matching

Choose the most appropriate description for each ecological term. [p.462]

1. ____ ecology
2. ____ habitat
3. ____ community
4. ____ niche
5. ____ specialist species
6. ____ generalist species
7. ____ ecosystem
8. ____ biotic
9. ____ succession
10. ____ biomes

A. In any given habitat, the populations of all species that associate directly or indirectly

B. A community of organisms interacting with each other and with the physical environment

C. Consists of all the physical, chemical and biological conditions a species requires to live and reproduce

D. An orderly progression of species replacement in new or disturbed habitats

E. The type of place a species normally lives; its home

F. Species with broad niches that live in more than one habitat and eat many types of food

G. Species with narrow niches that have specific requirements for habitat and food choices

H. The living portion of an ecosystem

I. The study of the interactions or organisms with each other and with the physical environment

J. Major land areas with differing types of dominant vegetation

Choice

For questions 11-14, choose from the following:

   a. primary succession [p.464]   b. secondary succession [p.464]

11. ____   Successional changes begin when a pioneer species colonizes a newly available habitat.

12. ____   A community develops toward the climax state after parts of the habitat have been disturbed.

13. ____   This pattern occurs in abandoned fields in which wild grasses and other plants quickly take hold when cultivation stops.

14. ____   Might occur on a recently deglaciated area.

24.2. FEEDING LEVELS AND FOOD WEBS [pp.466-467]

24.3. ENERGY FLOW THROUGH ECOSYSTEMS [p.468]

Boldfaced Terms

producers_____

_____

consumers _____

_____

food chain _____

_____

food webs _____

_____

primary productivity _____

_____

ecological pyramid _____

_____

Fill-in-the-Blanks [p.466]

Nearly every ecosystem runs on energy from the sun that is captured by (1) _____, plants and

other photosynthetic organisms, known as (2) _____ or "self-feeders". All other organisms in the

ecosystem are (3) _____ that depend directly or indirectly on energy stored in the tissues of producers.

Consumers are "other-feeders" or (4) _____ . Within this group are (5) _____ like mice,

rabbits and insects that eat plants, (6) _____ like hawks and snakes that eat animals, (7)

_____ like humans and bears that feed on both plant and animal foods, and (8) _____ like fungi, bacteria and worms that get energy and nutrients from the remains or products of organisms. Producers obtain an ecosystem's nutrients and its initial pool of (9) _____ . Nutrients are passed to other members of the ecosystem and eventually broken down and (10) _____ by decomposers for re-use by producers.

Choice

For questions 11-18, choose from the following [p.466]:

　　a. primary producer　　b. herbivore　　c. primary carnivore　　d. secondary carnivore
　　e. decomposer　　　　f. omnivore

11. _____　gain energy directly from sunlight
12. _____　a hawk that eats a snake
13. _____　fungi and bacteria
14. _____　organisms that prey on herbivores
15. _____　humans and others who consume both plant and animal material
16. _____　green plants
17. _____　autotrophs
18. _____　snails, grasshoppers, and other plant-eaters

Labeling

Identify the levels of the following diagram of an energy pyramid. [p.468]

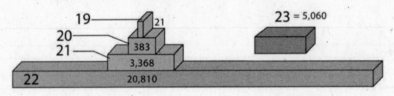

19. _____

20. _____

21. _____

22. _____

23. _____

Dichotomous Choice

Circle one of the two possible answers given between parentheses in each statement. [pp.466-468]

24. Minerals carried by erosion into a lake represent nutrient (input / output) to the lake ecosystem.

25. Ecosystems require a continual energy input from the (sun / heterotrophs).

26. Nutrients typically (are / are not) recycled.

27. Each species in an ecosystem fits somewhere in a hierarchy of feeding relationships called (niche / trophic) levels.

28. A linear sequence of who eats whom in an ecosystem is sometimes called a food (chain / web).

29. Food chains cross-connect with one another in (food webs / biomes).

30. The amount of energy actually stored in the primary trophic level of an ecosystem depends on how many (plants / animals) are present, and on the balance between energy trapped by photosynthesis and energy used by the plants.

31. In a harsh ecosystem environment, productivity would be expected to be (lower / higher) than in a mild ecosystem environment.

32. In an ecological pyramid, the primary (producers / consumers) form a base for successive tiers of (producers / consumers) above them.

33. A(n) (biomass / energy) pyramid depicts the weight of all an ecosystem's organisms.

34. In an energy pyramid, the amount of available energy (increases / decreases) through successive feeding levels of an ecosystem.

35. The rate at which primary producers store energy in their tissues is the ecosystem's (rate of photosynthesis / primary productivity).

24.4. INTRODUCTION TO BIOGEOCHEMICAL CYCLES [p.469]

24.5. THE CARBON CYCLE [p.470-471]

Boldfaced Terms

biogeochemical cycle _____

_____

water cycle _____

_____

carbon cycle_____

_____

Fill-in-the-blanks [p.469]

The availability of water, carbon dioxide and minerals that serve as nutrients for (1) _____ have a major impact on ecosystems. Ions and molecules of nutrients move back and forth between the environment and organisms in (2) _____ cycles . A part of the environment serves as a storage area, or (3) _____ , for individual nutrients. Nutrients move (4) _____ between organisms and the environment, but (5) _____ through a reservoir.

There are three basic types of biogeochemical cycles. In the (6) _____ cycle, oxygen and hydrogen move in the form of water molecules. Much of the nutrient is in the form of a gas in (7) _____ cycles. (8) _____ cycles move solid nutrients from land to seafloor then back to dry land through the process of (9) _____ .

Matching

Choose the description that best matches the numbered term. [p.469]

10. ____ solar energy
11. ____ ocean
12. ____ forms of precipitation
13. ____ ocean currents and prevailing winds
14. ____ forms of atmospheric water
15. ____ geologic uplift

A. Mostly rain and snow

B. Affect movement of water in global patterns

C. Water vapor, clouds, and ice crystals

D. As crustal plates move, minerals stored in seafloor sediments become available to land organisms

E. Slowly drives water from ocean to atmosphere to land and back to the ocean

F. Main reservoir of water

Labeling Diagrams

Use the following illustration of the hydrologic cycle to answer questions 16-21. [p.469]

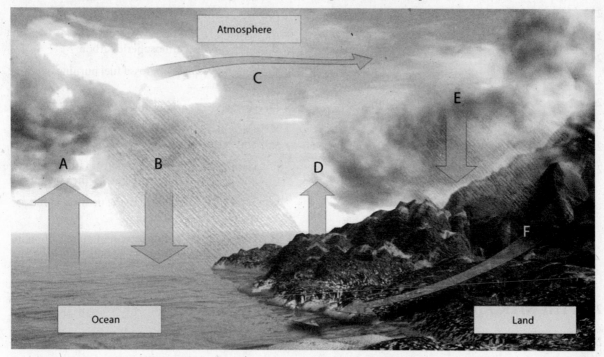

16. What does the arrow at 'A' represent? _____

17. What does the arrow at 'B' represent? _____

18. What does the arrow at 'C' represent? _____

19. What does the arrow at 'D' represent? _____

20. What does the arrow at 'E' represent? _____

21. What does the arrow at 'F' represent? _____

Principles of Ecology   373

Matching

Choose the description that best matches the numbered term concerning the carbon cycle. [pp.470-471]

22. _____ Peat
23. _____ Bicarbonate and carbonate
24. _____ Ocean reservoir
25. _____ Carbon cycle
26. _____ Sources of atmospheric carbon
27. _____ Carbon dioxide (CO_2)
28. _____ Climate change and global warming
29. _____ Fossil fuels

A. Deep storage of CO_2; affects global carbon budget

B. Form of most atmospheric carbon

C. Aerobic respiration, fossil fuel burning, volcanic eruptions

D. A reserve of carbon stored in anaerobic bogs

E. Carbon trapped in the form of natural gas, coal and petroleum

F. Carbon moves from reservoirs in the atmosphere and oceans, through organisms, then back again

G. Most carbon dissolved in the ocean is in these forms

H. Caused by the accumulation of more carbon in the atmosphere than can be returned to the ocean reservoir

24.6. THE NITROGEN CYCLE [p.472]

Boldfaced Terms

nitrogen cycle _____

_____

nitrogen fixation _____

_____

Choice

For questions 1-7, choose from the following [p.472]

    a. nitrogen cycle   b. nitrogen fixation   c. nitrogen   d. nitrification   e. denitrification

1. _____ Ammonia and ammonium in soil are converted to nitrite (NO_2); other bacteria metabolize nitrite and convert nitrite to nitrate (NO_3).

2. _____ Bacteria are its key organisms—they convert nitrogen to forms plants can use, and also release nitrogen to complete the cycle.

3. _____ Component of amino acids, proteins and nucleic acids.

4. _____ Process in which a few kinds of bacteria convert N_2 to ammonia (NH_3), which dissolves quickly in water to form ammonium (NH_4^+).

5. _____ Gaseous form makes up 80 percent of the atmosphere.

6. _____ Plants called legumes have mutually beneficial associations with bacteria that carry out this process.

7. _____ Bacteria convert nitrate or nitrite to N_2 (and a bit of nitrous oxide) that escapes into the atmosphere.

Labeling

Label the parts of the nitrogen cycle based on the illustration below. [p.472]

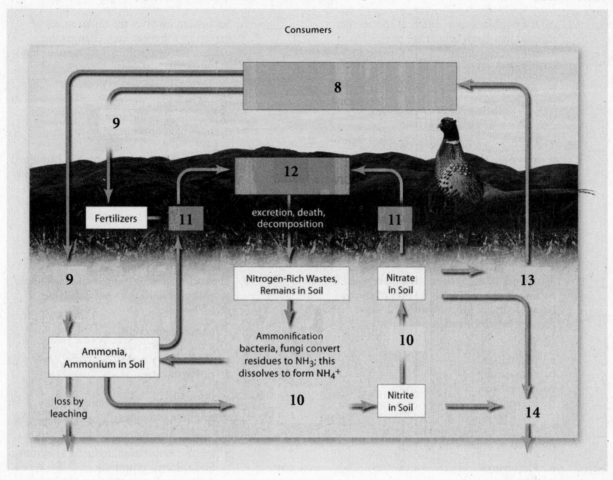

8. _____

9. _____

10. _____

11. _____

12. _____

13. _____

14. _____

SELF-QUIZ

___ 1. All the populations of different species that interact in a given habitat are referred to as a(n)_____. [p.464]

    a. biosphere
    b. community
    c. ecosystem
    d. niche

___ 2. The _____ of a species consists of all the physical, chemical, and biological conditions that the species needs to live and reproduce in an ecosystem. [p.464]

    a. habitat
    b. niche
    c. carrying capacity
    d. ecosystem

___ 3. A network of interactions that involve the cycling of materials and the flow of energy between one or more communities and the physical environment is a(n) _____. [p.464]

    a. population
    b. community
    c. ecosystem
    d. biosphere

___ 4. _____ get energy from the remains or products of organisms. [p.466]

    a. Herbivores
    b. Parasites
    c. Decomposers
    d. Carnivores

___ 5. A linear sequence of who eats whom in an ecosystem is sometimes called a(n)_____. [p.467]

    a. trophic level
    b. food chain
    c. ecological pyramid
    d. food web

___ 6. In a typical food web, the second trophic level includes primary consumers that are _____. [p.467]

    a. herbivores
    b. carnivores
    c. scavengers
    d. decomposers

___ 7. Succession involves the replacement of species by others as each changes the habitat, finally resulting in a more or less stable _____. [p.464]

    a. climax community
    b. pioneer community
    c. secondary community
    d. pyramid community

___ 8. A (an) ____ ____ represents the combined weights of all organisms at each level of an ecosystem. [p.468]

    a. food chain
    b. food web
    c. productivity map
    d. biomass pyramid

___ 9. The cycling of phosphorous is an example of what type of biogeochemical cycle? [p.469]

    a. global water cycle
    b. atmospheric cycle
    c. sedimentary cycle
    d. nitrogen cycle

___ 10. In the carbon cycle, carbon enters the atmosphere through _____. [p.470]

    a. carbon dioxide fixation
    b. aerobic respiration, fossil fuel burning, and volcanic eruptions
    c. oceans and accumulation of plant biomass
    d. evaporation

___ 11. Which of the following *is not true* about the water cycle? [p.469]

    a. water returns to the atmosphere via evaporation from oceans and land
    b. water returns to the oceans and land via precipitation from the atmosphere
    c. water returns to the atmosphere via transpiration of plants
    d. surface and groundwater flow brings water from oceans to land

_____ 12. Which type of nutrient cycling is linked to changes in climate? [p.470]

    a. water cycle
    b. phosphorus cycle
    c. carbon cycle
    d. nitrogen cycle

_____ 13. _____ is the name for the process in which certain bacteria convert N2 to ammonia which dissolves to form ammonium, a form that plants can pull into an ecosystem. [p.472]

    a. nitrification
    b. ammonification
    c. denitrification
    d. nitrogen fixation

_____ 14. Legumes and their mutualistic bacteria are associated with which part of the nitrogen cycle? [p.472]

    a. nitrification
    b. ammonification
    c. denitrification
    d. nitrogen fixation

CHAPTER OBJECTIVES/REVIEW QUESTIONS

1. Define and distinguish between habitat and niche. [p.464]

2. What does the science of ecology encompass? [p.464]

3. What distinguishes a community from an ecosystem? [p.464]

4. Distinguish between primary succession and secondary succession. [p.464]

5. Name and distinguish among the four types of consumers in food webs. [p.466]

6. Distinguish between a food chain and a food web. Which is most often found in ecosystems? [pp.466-467]

7. Define primary productivity. [p.468]

8. Biomass pyramids represent the _____ of the entire ecosystem's members at each trophic level; _____ pyramids reflect the energy losses as energy flows through the trophic levels of an ecosystem. [p.468]

9. Name the three types of biogeochemical cycles. [p.469]

10. In a _____ cycle, minerals moves from land to sediments in the seas, and then back to the land. [p.469]

11. Explain the functions of evaporation, transpiration, precipitation and surface and groundwater flow in the hydrologic cycle. [p.469]

12. What contributes to changes in climate? With which cycle is this process linked? [pp.470-471]

13. What might be some of the fates of a carbon dioxide molecule after it is incorporated into glucose during photosynthesis? [pp.470-471]

14. Define the chemical events that occur during nitrogen fixation, nitrification, ammonification, and denitrification. [p.472]

INTEGRATING AND APPLYING KEY CONCEPTS

1. Is there *a fundamental niche* that is occupied by humans? If you think so, describe the minimal abiotic and biotic conditions required by populations of humans in order to live and reproduce. (Note that "to be happy", "popular" or "financially successful" are not requirements for survival.)

2. In a food web, several different species may feed at the same level thereby introducing competition. What do you think might happen if someone purchases a tropical fish from South America and releases it into a lake in the southern United States where the climate is similar to its home?

25

HUMAN IMPACTS ON THE BIOSPHERE

INTRODUCTION

In Chapter 24 you were introduced to the science of ecology. You will now apply the knowledge you gained concerning populations, communities, ecosystems, habitats, niches, trophic levels and nutrient cycling. Evolution sculpted the biodiversity now on earth including our species, *Homo sapiens*. We are now threatening to change the ecological balance and diversity that took so long to achieve. This chapter examines how the growing population of humans has changed, and is continuing the change, planet earth. You will learn what effects our activities have on the biosphere, the survival of other species, our own continued existence, and what can be done to stop further destruction of the earth, our home and habitat.

FOCAL POINTS

- Many environmental problems the world is facing can be traced to human population growth and demand for natural resources. [p.479]
- Figure 25.1 [p.476] is a good graphic demonstration of the growth of the human population throughout history.
- Figure 25.4 [p.478] illustrates the S-shaped curve typical of a logistic growth pattern.
- Figure 25.6 [p.480] shows normal air mass layering over cities and the layering pattern during a thermal inversion resulting in smog.
- Figure 25.8 [p.481] diagrams the stratification of the atmosphere with particular emphasis on the ozone layer.
- Figure 25.9 [p.482] and figure 25.10 [pp.482-483] demonstrate the greenhouse effect and graphically depict changes in global temperature in the last 133 years.
- Figure 25.19 [p.488] crosslinks the sources of energy used in the United States with the consumers of that energy

INTERACTIVE EXERCISES

CHAPTER INTRODUCTION [p.476]

25.1. HUMAN POPULATION GROWTH [pp.476-477]

25.2. NATURE'S CONTROLS ON POPULATION GROWTH [p.478]

Boldfaced Terms

total fertility rate (TFR) _____

_____

population density _____

_____

age structure _____

_____

limiting factor _____

_____

carrying capacity _____

_____

Dichotomous Choice

Circle one of the two possible answers given between parentheses in each statement. [pp.476-478]

1. The world population reached 7 (million / billion) in 2011.

2. (Growth rate / fertility rate) is determined by the balance between births, deaths, immigration and emigration.

3. Overall on Earth, (birth / death) rates have been coming down; (birth / death) rates have been falling, too.

4. The countries expected to show the most population growth are China and (the United States / India).

5. Currently, the total global fertility rate is (below / above) replacement level fertility.

6. TFRs are at or below replacement levels for many (developing / developed) countries.

7. Even if every couple decides to bear no more than two children, the world population will keep growing for (twenty / sixty) years.

8. A population's vital statistics—size, age structure, density—are its (national base / demographics).

9. The total number of individuals in a given area is population (density / distribution).

10. Dividing a population into prereproductive, reproductive, and postreproductive categories characterizes its (age / distribution) structure.

11. The reproductive (rate / base) of a population refers to the number of individuals in the prereproductive and reproductive age structure categories.

12. The population of the United States has a narrow reproductive base and exhibits (slow / rapid/ growth).

13. Populations experiencing exponential growth (double / triple) with each generation.

14. The human population has been experiencing (exponential / logistic) growth, since the mid-1700s.

15. When the course of logistic growth is plotted on a graph, a(n) (J-shaped / S-shaped) curve is obtained.

16. Earth is already exceeding, or about to exceed, its (density / carrying capacity) for humans.

17. Infectious diseases and parasites are a bigger problem in (low-density / high-density) populations.

Matching

Select the best description for each term. [p.478]

18. ____ density-independent controls
19. ____ carrying capacity
20. ____ limiting factor
21. ____ logistic growth
22. ____ density-dependent controls
23. ____ biotic potential
24. ____ exponential growth

A. A growth pattern in which a population grows slowly at first, then increases rapidly, but levels off when carrying capacity is reached

B. An essential resource such as food and water that, when in short supply, affects population growth

C. The maximum reproductive ability of a population under ideal conditions

D. An example is overcrowding resulting in competition for resources

E. Events such as natural disasters that cause deaths or births regardless of whether crowding exists

F. The number of individuals of a given species that can be sustained indefinitely by the resources in a given area

G. Population growth in doubling increments

25.3. ECOLOGICAL "FOOTPRINTS" AND ENVIRONMENTAL PROBLEMS [p.479]

Boldfaced Terms

ecological footprint _____

_____

renewable resources _____

_____

nonrenewable resources _____

_____

pollutant _____

_____

Fill-in-the-Blanks [p.479]

(1) _____ and the unsustainable use of (2) _____ are the two major causes of today's environmental problems. A population's, or individual's, (3) _____ encompasses both the total resources it consumes and the wastes is produces. Ecological footprints vary. An individual in the (4) _____ consumes 100 times more resources than a person in very poor areas of Africa or Asia.

Water supply and forests can be replenished, therefore they are considered (5) _____ . The earth's crust contains limited amounts of fossil fuels and minerals. Therefore they are considered (6) _____ .

Wastes and chemical by-products of industry and agriculture create (7) _____ . A (an) (8) _____ in some way harms the health or survival of a population. Most pollutants come from (9) _____ activities. When pollutants come from a single place, like a leaky drum of waste, they come from a (an) (10) _____ . Pollutants coming from a widespread area, like pesticide runoff from homes and farms, come from a (an) (11) _____ .

Dichotomous Choice

From each of the following pairs of terms choose the term which leaves the smallest ecological footprint. [p.479]

12. Riding a bicycle / driving a car
13. Using an air conditioner / using a fan
14. Cutting down evergreens for Christmas trees and replanting with new trees / logging a rainforest to build houses
15. Buying vegetables from a supermarket that have been grown far away / growing a home vegetable garden

25.4. ASSAULTS ON OUR AIR [pp.480-481]

Boldfaced Terms

smog _____

_____

acid rain _____

_____

ozone thinning _____

_____

chloroluorocarbons (CFCs) _____

_____

Choice

For questions 1-12, choose from the following aspects of atmospheric pollution. [pp.480-481]

  a. chlorofluorocarbons   b. industrial smog   c. photochemical smog   d. acid rain   e. ozone layer

1. _____ Develops as a brown, smelly haze over large cities.

2. _____ Precipitation containing weak sulfuric and nitric acids.

3. _____ Contributes to ozone reduction more than any other factor.

4. _____ Where winters are cold and wet, this develops as a gray haze over many cities that burn coal and other fossil fuels.

5. _____ Substances from power plants and factories as well as from motor vehicles combine with water producing a solution with a pH as low as 2.3.

6. _____ Each year, from September through mid-October, it thins down by as much as half above Antarctica.

7. _____ Today most of this forms in cities of China, India, eastern Europe, and other developing countries.

8. _____ Contains airborne pollutants, including dust, smoke, soot, ashes, asbestos, oil, bits of lead and other heavy metals and sulfur dioxide.

9. _____ Reduction allows more harmful ultraviolet radiation to reach the earth's surface.

10. _____ Nitric oxide released from vehicles is exposed to sunlight and reacts with hydrocarbons to form harmful oxidants.

11. _____ Widely used as refrigerator coolants, air conditioners, industrial solvents, and plastic foams; enter the atmosphere slowly and resist breakdown.

12. _____ Protects against skin cancer and cataracts.

25.5. GLOBAL WARMING AND CLIMATE CHANGE [pp.482-483]

Boldfaced Terms

greenhouse effect _____

_____

global warming _____

_____

global climate change _____

_____

Fill-in-the-Blanks [pp.482-483]

Atmospheric gases like carbon dioxide, water, ozone, methane, nitrous oxide and CFCs that affect global

temperature are collectively called (1) _____ . These gases slow the escape of (2) _____ back

into space. The (3) _____ refers to the warming of Earth's surface by heat energy trapped in the (4)

_____ atmosphere by greenhouse gases. Carbon dioxide levels in the atmosphere of the Northern

Human Impacts on the Biosphere   **383**

Hemisphere fluctuate naturally in an annual cycle along with the amount of (5) _____ of autotrophs. Carbon dioxide levels are lowest when the rate of (6) _____ is highest.

Atmospheric levels of greenhouse gases are far (7) _____ than in the past. Carbon dioxide may be at its highest level since (8) _____ years ago. Most climate scientists agree that the increase in greenhouse gases is a factor in (9) _____ . The temperature of the lower atmosphere has risen (10) _____ degree Fahrenheit in a little over a century. Irreversible (11) _____ is happening at an ever increasing pace.

Continued temperature increases will affect the climate of every place on earth. Some changes may appear locally beneficial, others will be disastrous. Patterns of (12) _____ are changing; some areas flood while others experience (13) _____ . Polar ice is (14) _____ and (15) _____ are retreating. As a result (16) _____ will rise in many coastal areas. The IPCC predicts that by (17) _____ , the seas may submerge up to a third of the world's coastal wetlands and coral reefs. Large chunks of (18) _____ like the Atlantic Coast of the United States may erode away.

25.6. PROBLEMS WITH WATER AND WASTES [pp.484-485]

25.7. PROBLEMS WITH LAND USE AND DEFORESTATION [pp.486-487]

Boldfaced Terms

desertification _____

_____

deforestation _____

_____

Matching

Select the description which best matches the numbered term. [pp.484-487]

1. ____   groundwater
2. ____   salinization
3. ____   saltwater intrusion
4. ____   landfill
5. ____   NIMBY
6. ____   incinerators
7. ____   forested watersheds
8. ____   green revolution

9. ____   desertification
10. ____   deforestation

A. Act like giant sponges that absorb, hold and gradually release water

B. Community mindset: "not in my backyard"

C. Overgrazing of livestock on marginal lands is a prime cause of this phenomenon

D. Being depleted by large-scale irrigation as in the Ogallala aquifer

E. Overuse of groundwater causes seawater to invade the water supply in this process

F. Removal of trees from large areas of land for logging, agriculture or herd grazing, causes erosion and nutrient loss

G. A place humans bury solid waste; they leak and threaten to contaminate groundwater supplies

H. Mineral salt buildup in poorly drained soils affects the growth of crops

I. Burning of solid wastes can produce a toxic ash remainder

J. A principle based on the use of fertilizers, pesticides and irrigation to grow high-yield crops

25.8. MOVING TOWARD RENEWABLE ENERGY SOURCES [pp.488-489]

Boldfaced Terms

fossil fuels _____

_____

biofuels _____

_____

Crossword Puzzle

Solve the crossword puzzle using the following clues. [pp.488-489]

ACROSS

a. A renewable source of energy from the sun [p.489]
f. A renewable source of energy made from plants and organic wastes [p.489]
h. A nonrenewable source of energy from the remains of ancient forests [p.488]
i. A type of vehicle that runs on both gasoline and electric power [p.489]
j. A type of biofuel made mainly from corn [p.489]
l. A power corridor of this type of energy stretching from Texas to the Dakotas could meet the electricity needs of 80% of the continental U.S. [p.489]
m. A type of fossil fuel [p.488]
n. A danger of nuclear power is a meltdown of the reactor _____. [p.489]
o. This type of fallout is another danger of nuclear power plants. [p.489]

DOWN

b. _____ cells that produce electricity are laid out in panels to collect sunlight. [p.489]
c. There has been much research aimed at developing _____ _____ cells to power cars. [p.489]
d. A type of fossil fuel [p.488]
e. Burning of fossil fuels adds to the _____ effect. [p.488]
g. Giant engines turned by the power of the wind [p.489]
k. A renewable source of energy that is inexpensive and clean but raises safety concerns [p.489]

Interpreting Diagrams

Follow the intersecting lines on the diagram below to answer questions 2 – 7 concerning energy usage in the United States. [p.488]

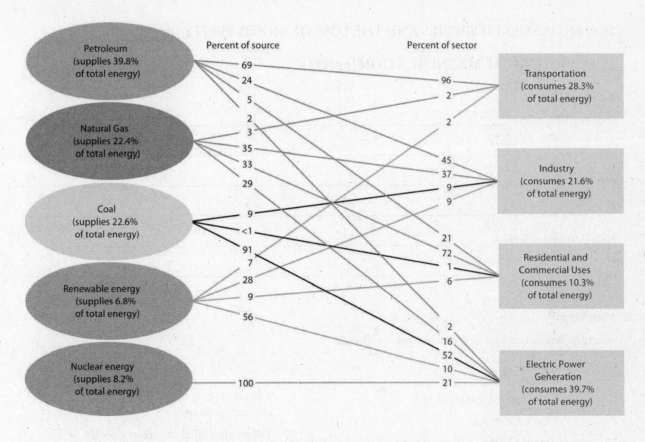

Percent of source

Petroleum (supplies 39.8% of total energy)
- 69
- 24
- 5

Natural Gas (supplies 22.4% of total energy)
- 2
- 3
- 35
- 33
- 29

Coal (supplies 22.6% of total energy)
- 9
- <1
- 91

Renewable energy (supplies 6.8% of total energy)
- 7
- 28
- 9
- 56

Nuclear energy (supplies 8.2% of total energy)
- 100

Percent of sector

Transportation (consumes 28.3% of total energy)
- 96
- 2
- 2

Industry (consumes 21.6% of total energy)
- 45
- 37
- 9
- 9

Residential and Commercial Uses (consumes 10.3% of total energy)
- 21
- 72
- 1
- 6

Electric Power Generation (consumes 39.7% of total energy)
- 2
- 16
- 52
- 10
- 21

2. Ninety-six percent of the energy consumed by transportation comes from what source? _____

3. One-hundred percent of nuclear energy is used for this purpose. _____

4. Most electric power generated in the U.S. comes from what source? _____

5. *Residential and Commercial Uses'* main source of energy is _____

6. Most of the energy consumed by *Industry* comes from _____

7. Assume that suddenly there is a very limited supply of fossil fuels. List the four sectors of energy in order of impact. List the sector that would be impacted the most first, list the next impacted sector second, etc, ending with the sector that would be impacted least. _____

_____

_____

_____

25.9. ENDANGERED SPECIES AND THE LOSS OF BIODIVERSITY [pp.490-491]

25.10. BIOLOGICAL MAGNIFICATION [pp.491]

Boldfaced Terms

endangered species _____

_____

sustainability _____

_____

biological magnification _____

_____

Matching

Select the description that best matches the numbered term. [pp.490-491]

1. ____ endangered species
2. ____ endemic species
3. ____ tropical deforestation
4. ____ rosy periwinkle
5. ____ habitat loss
6. ____ biological magnification

A. A tropical plant that is the source of two major anti-cancer drugs

B. The greatest killer of species globally

C. A species extremely vulnerable to extinction

D. The physical reduction of suitable living spaces for wildlife

F. An increase in the concentration of a nondegradable (or slowly degradable) substance in the tissues of organisms as it is passed upwards through food chains

I. A species that originated in one geographic region and lives nowhere else

Fill-in-the-Blanks [p.491]

Wildlife can be harmed and even driven to the brink of extinction by the process of (7) _____ .

During World War II, DDT was sprayed in the tropical Pacific to control (8) _____ responsible for transmitting the organisms that cause dangerous malaria. Later in Europe, it was used to control (9) _____ that transmitted a bacterium that causes typhus. After the war, the use of DDT continued to be used against agricultural, garden or forest pests. Many animals began dying. DDT can accumulate in the (10) _____ of animals that come into contact with it. Since DDT breaks down slowly, it is passed from (11) _____ (on whom it is sprayed) to (12) _____ up through the food chain. DDT

also killed off natural (13) _____ that had been keeping the pest population in check. The most

devastated species were at the (14) _____ of the food chain. Large birds of prey, like eagles, ospreys

and falcons began producing eggs with very thin (15) _____ due to the DDT metabolism. Many chick

embryos died, and some species were facing (16) _____ .

SELF-QUIZ

____ 1. _____ refers to an increase in concentration of a nondegradable (or slowly degradable) substance in organisms as it is passed along food chains. [p.491]

    a. Ecosystem modeling
    b. Nutrient input
    c. Biogeochemical cycle
    d. Biological magnification

____ 2. The average number of individuals of the same species per unit area at a given time is the _____. [p.477]

    a. population density
    b. population growth
    c. population birth rate
    d. population size

____ 3. The maximum number of individuals of a population (or species) that can be sustained by a given environment defines _____. [p.478]

    a. the carrying capacity
    b. exponential growth
    c. logistic growth
    d. density-independent factors

____ 4. Which of the following is *not* characteristic of logistic growth? [p.478]

    a. S-shaped curve
    b. growth levels off as carrying capacity is reached
    c. unrestricted growth
    d. slow growth of a low-density population followed by rapid growth

____ 5. Factors that are at work in crowded populations, and tend to reduce population size, are called _____. [p.478]

    a. density-independent controls
    b. density-dependent controls
    c. limiting factors
    d. logistic factors

____ 6. _____ like floods, earthquakes and natural disasters can also reduce population size. [p.478]

    a. Density-independent controls
    b. Density-dependent controls
    c. Limiting factors
    d. Logistic factors

____ 7. Resources such as minerals and fossil fuels are _____. [p.479]

    a. point sources
    b. nonpoint sources
    c. renewable resources
    d. nonrenewable resources

____ 8. _____ result(s) when nitrogen dioxide and hydrocarbons react in the presence of sunlight. [p.480]

    a. Photochemical smog
    b. Industrial smog
    c. A thermal inversion
    d. Eutrophication

____ 9. Sulfur and nitrogen dioxides dissolve in atmospheric water to form a weak solution of sulfuric acid and nitric acid; this describes the formation of _____. [p.480]

    a. photochemical smog
    b. industrial smog
    c. ozone and CFCs
    d. acid rain

____ 10. Which of the following is *not* true about ozone? [p.481]

    a. a layer of ozone exists between the stratosphere and the mesosphere
    b. ozone is a molecule of three oxygen atoms
    c. ozone absorbs UV radiation and protects against skin cancer and cataracts
    d. CFCs deplete ozone

11. Which of the following *does not* contribute to the greenhouse effect? [p.482]

    a. carbon dioxide released by burning fossil fuels
    b. the presence of CFCs in the atmosphere
    c. acid rain
    d. greenhouse gases

12. Which of the following does not pose a threat to the availability of a supply of fresh water? [p.484-485]

    a. salinization
    b. saltwater intrusion
    c. landfills
    d. forested watersheds

13. _____ is the removal of all trees from large tracts of land for logging, agricultural or grazing purposes. [p.487]

    a. Salinization
    b. Desertification
    c. Deforestation
    d. Stripping

14. Which of the following energy sources is renewable? [p.488-489]

    a. biodiesel
    b. petroleum
    c. natural gas
    d. coal

15. Endangered species are threatened by _____. [p.490]

    a. habitat loss
    b. overharvesting
    c. pollution
    d. all of the above

CHAPTER OBJECTIVES/REVIEW QUESTIONS

1. Discuss factors that determine the growth rate of a population, both those that increase it and those that cause it to decrease. [p.476]

2. Define the three stages of a population's age structure. [p.477]

3. Distinguish between exponential growth and logistic growth. [p.478]

4. Define limiting factors and tell how they influence carrying capacity. [p.478]

5. Define *density-dependent controls,* give two examples, and indicate how density-dependent factors act on populations. [p.478]

6. Define *density-independent controls* and give two examples. [p.478]

7. Define "ecological footprint." [p.479]

8. Distinguish between renewable and nonrenewable resources. [p.479]

9. What is the difference between a point source and a nonpoint source of pollution? [p.479]

10. Identify the principal air pollutants, their sources, and their effects. [p.480]

11. Distinguish between industrial smog and photochemical smog. [p.480]

12. Describe the environmental conditions leading to acid rain, and list its negative effects. [p.480]

13. What is the ozone layer, what advantage does it provide to the earth's inhabitants, and what specific factors are threatening its existence? [p.481]

14. Explain the greenhouse effect and list the gases that contribute to it. What are the implications that global warming has for life on Earth? [pp.482-483]

15. Summarize the major problems contributing to contamination of the world's supply of pure water. [p.484-485]

16. Describe the process of desertification and its effects. [p.486]

17. Explain the repercussions of deforestation. [p.487]

18. List the disadvantages of the use of fossil fuels as energy sources. [p.488]

19. List some examples of alternative energy sources. Discuss their pros and cons. [p.489]

20. Cite major reasons for the continuing loss of biodiversity. [p.490]

21. Distinguish between endangered species and endemic species. [p.490]

22. Explain the detrimental effects of biological magnification. [p.491]

23. What is the principle of sustainability? [p.491]

INTEGRATING AND APPLYING KEY CONCEPTS

1. Considering the rapid rate of species and habitat loss in the 21st century, why do you think some biologists feel that the endangered species list should include all species?

2. How would the global ecological footprint be affected if the total fertility rate dropped to below replacement level?

3. Do you think the human race has an ethical right to "increase and multiply" to the exclusion of other species on earth? In the giant picture of things, are we the penultimate of evolution and therefore entitled to own the entire earth? What challenges will face us when we overreach our carrying capacity? How will the loss of other species affect us in the long run?

4. Consider the following statement: "The development of technology has enabled human populations to overcome density-dependent controls and has led to the population explosion we now see." Do you think this is true? If so, what effects will the human population feel as it continues to grow exponentially?

Answer Key

1 LEARNING ABOUT HUMAN BIOLOGY
Chapter Introduction [p.1]
1.1. The Characteristics of Life [p.2]
1. a, 2. c, 3. b, 4. d, 5. c, 6. e, 7. c, 8. d, 9. e, 10. b, 11. e, 12. d, 13. c, 14. a
1.2 Our Place In The Natural World [p.3]
1. animal, 2. vertebrates, 3. mammals, 4. primates, 5. humans, 6. protists, fungi, plants, animals, 7. Archaea, Bacteria
1.3 Life's Organization [pp.4-5]
1. d, 2. c, 3. e, 4. f, 5. g, 6. i, 7. j, 8. b, 9. h, 10. a, 11. e, 12. k, 13. i, 14. j, 15. c, 16. a, 17. f, 18. g, 19. h, 20. d, 21. b, 22. sun, 23. producers, 24. consumers, 25. decomposing, 26. webs
1.4. Using Science to Learn About the Natural World [pp.6-7]
1.5. Critical Thinking in Science and Life [p.8]
1.6. Science in Perspective [p.9]
1.a. observation, b. hypothesis, c. prediction, d. experiment, e. variable, f. control group, 2. f, 3. a, 4. d, 5. b, 6. c, 7. e, 8. A fact consists of information that can be verified. An opinion cannot be verified because it consists of a subjective judgment., 9. Gather information from reliable sources; separate fact from opinion; use facts that can be verified independent of the source; don't confuse cause and correlation., 10. A hypothesis is a proposed explanation for a question generated by observation. Hypotheses are tested by experimentation. A theory is the explanation of a broad range of observations based on repeated testing of related hypotheses., 11. false, 12. false, 13. true, 14. true, 15. true, 16. true, 17. false, 18. is correlated to, 19. critical, 20. opinion
1.7. Living in a World of Infectious Disease [pp.10-11]
1.8. Homeostasis [p.11]
1. f, 2. e, 3. b, 4. i, 5. j, 6. c, 7. d, 8. a, 9. g, 10. h, 11. false, 12. false, 13. true, 14. true

Self-Quiz
1. c, 2. d, 3. d, 4. c, 5. a, 6. b, 7. c, 8. b, 9. a, 10.d, 11. c, 12. d, 13. c

---

2 CHEMISTRY OF LIFE
Chapter Introduction [p. 15]
2.1 Atoms and Elements [p.16-17]
2.2 Pet Scanning – Using Radioisotopes in Medicine [p.17]
2.3 Chemical Bonds: How Atoms Interact [p.18-19]
1. a, 2. b, 3. j, 4.k, 5. l, 6. n, 7. c, 8. m, 9. d, 10. o, 11. p, 12. g, 13. I, 14. e, 15. f, 16. h, 17. mass number, 18.

mass number, 19. less, 20. outer, 21. Oxygen and water are both molecules because they are a combination of two or more atoms bonded together. Water is also a compound because it is a molecule that contains atoms of two elements whereas oxygen gas contains two atoms of only one element. 22. energy level, 23. mixture, 24. calcium, 25. chemical equations, 26. reactants, 27. products, 28. balanced.
2.4. Important Bonds in Biological Molecules [pp.20–21]
2.5. Water: Necessary for Life [pp.22-23]
1. A weak attraction forms between a covalently bound hydrogen atom and an electronegative atom in another molecule. The bond is basically an attractive force similar to that which attracts the opposite poles of two magnets., 2. The outer shell of a hydrogen atom contains only one electron leaving room for only one electron from another atom to share the space. The outer shell of an oxygen atom has six electrons leaving room for two electrons from another atom to share the space. The outer shell of a nitrogen atom has five electrons leaving room for the sharing of three electrons from another atom.
3.

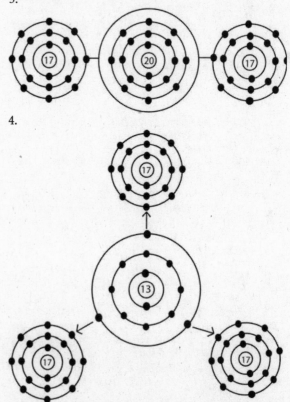

4.

5. donated, accepted, 6. opposite, 7. covalent, polar, 8. covalent, 9. hydrogen, 10. blood, 11. hydrogen, 12.

hydrophilic, 13. hydrophobic, 14. heat, 15. do not re-form, 16. losing, 17. solvent, 18. solutes, 19. dissolve.

2.6. How Antioxidants Protect Cells [p.23]

2.7. Acids, Bases, and Buffers: Body Fluids in Flux [pp.24-26]

1. Free radicals are created by oxidations that go on in the cells. This causes the free radical molecules to contain oxygen atoms that are lacking a full complement of electrons in the outer shell. This unstable state causes these atoms to "steal" an electron from another molecule causing disruption in structure and function of the affected molecule., 2. An antioxidant is a substance that gives up an electron to a free radical thus stabilizing it before it can do damage to DNA or some other vital cell component. Some antioxidants include vitamin C, vitamin E, and carotenoid which are pigments found in orange and leafy green vegetables., 3. acidic, 4. basic, 5. 7, 6. 100 times, 7. 100,000 times, 8. alkaline, 9. buffer, 10. H_2CO_3, HCO_3^-, 11. right, remove, 12. left, add, 13. acidosis, alkalosis, 14. coma, death, 15, acid, 16, base, 17. salt, 18. salt, 19. stays constant, 20. Milk of magnesia releases magnesium ions and hydroxide ions that combine with excess hydrogen ions in the stomach., 21. The buffer system accepts excess hydrogen ions thus raising the pH to acceptable levels., 22. When a strong base and a strong acid interact, a salt and water are formed., 23. Carbon dioxide combines with water to from carbonic acid and bicarbonate. If the blood pH rises, the carbonic acid will reverse the change. If the blood pH falls, the bicarbonate will reverse the change.; 24. a. 5, b. acid, c. 8, d. base, e. 2, f. acid, g. 1.3, h. acid, i. 9, j. base

2.8. Molecules of Life [pp.26–27]

1. true, 2. water, nucleic acids, 3. 2, 4, 4. true, 5. carbohydrate, hydrocarbon, 6. enzymes, functional group, 7. true, 8. rearrangement, functional groups transfer, 9. Enzymes act to speed up reactions., 10. Water is the byproduct of a condensation reaction. It is used to break apart monomers in a hydrolysis reaction. In a condensation reaction two or more monomers are bonded together and water is released. In a hydrolysis reaction two or more monomers are separated from each other by the addition of water., 11a. hydrolysis, 11b. yes.

2.9. Carbohydrates: Plentiful and Varied [pp.28–29]

1. a. sucrose, b. glucose, c. cellulose, d. deoxyribose, e. lactose, f. glycogen, g. starch

2.10. Lipids: Fats and Their Chemical Relatives [pp.30–31]

1a. saturated, 1b. unsaturated, 1c. polyunsaturated, 2. component of cell membranes, bile salts, estrogen, testosterone, 3. Two layers of phospholipid heads facing into the cell and outside of the cell and fatty acid tails facing each other in between., 4. b, 5. e, 6. d, 7. a, 8. c.

2.11. Proteins: Biological Molecules with Many Roles [pp.32–33]

2.12. A Protein's Shape and Function [pp.34–35]

1. amino group, 2. R group, 3. carboxyl group, 4. the R group,

5.

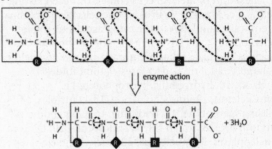

6. a, 7. c, 8. b, 9. d, 10. Glycoproteins have oligosaccharides boned to them while lipoproteins have cholesterol, triglycerides and phospholipids bound to them., 11. When a protein loses its normal three-dimensional shape it is denatured.

2.13. Nucleotides and Nucleic Acids [p.36]

2.14. Food Production and a Chemical Arms Race [p.37]

1. phosphate group, 2. sugar, 3. base, 4. The five carbon sugar in DNA is deoxyribose while the five carbon sugar in RNA is ribose.,

5. two complete nucleotides

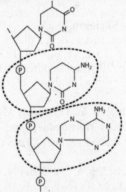

6. b, 7. a, 8. c, 9. Chemical agents may be used to protect crop yields, stored grains, ornamental plants and pets. The drawbacks include the killing of birds and other wildlife and the development of rashes, hives, headaches, asthma, joint pain, learning disabilities and behavioral problems.

Self-Quiz

1. a, 2. d, 3. c, 4. a, 5. c, 6. b, 7. c, 8. c, 9. b, 10. b

3 CELLS AND HOW THEY WORK

Chapter Introduction [p.41]

3.1. What Is a Cell? [pp.42-43]

3.2. Organelles of a Eukaryotic Cell [p.44]

1. All organisms have a least one cell. A cell is the smallest unit of life. All cells come from pre-existing cells., 2. c, 3. a, 4. b, 5. Prokaryotic cells have a plasma membrane and DNA in the cytoplasm. Eukaryotic cells contain their DNA in a membrane-bound nucleus and have other membrane-bound organelles. 6. large, 7. a lot of, 8. decreases, 9. thin; folded surfaces, 10. hydrophilic/hydrophobic, 11. cytoskeleton, 12. mitochondrion, 13. centrioles, 14. plasma membrane, 15. lysosome, 16. Golgi body, 17. smooth ER, 18. rough ER, 19. ribosomes, 20. DNA in nucleoplasm, 21. nucleolus, 22. nuclear envelope

3.3. How Do We See Cells? [p.45]

3.4. The Plasma Membrane: A Double Layer of Lipids [pp.46-47]

3.5. A Watery Disaster for Cells [p.47]

1. d, 2. e, 3. b, 4. a, 5. c, 6. phospholipids, 7. proteins, 8. c, 9. f, 10. b, 11. e, 12. d, 13. a, 14. mosaic, 15. phospholipids, 16. selective permeability, 17. nonpolar, 18. polar, 19. cholera, 20. exotoxin, 21. pump proteins, 22. small intestine, 23. water, 24. diarrhea

3.6. The Nucleus [pp.48-49]

3.7. The Endomembrane System [pp.50-51]

3.8. Mitochondria: The Cell's Energy Factories [p.52]

3.9. The Cell's Skeleton [p.53]

1. b, 2. c, 3. a, 4. e, 5. d, 6. d, 7. f, 8. e, 9. b, 10. c, 11. a, 12. 3, 13. 1, 14. 5, 15. 2, 16. 6, 17. 4, 18. "9+2 array", 19. cytoskeleton, 20. mitochondrion, 21. cytoskeleton, 22. mitochondrion, 23. centriole, 24. centriole,

3.10. How Diffusion And Osmosis Move Substances Across Membranes [pp.54-55]

3.11. Other Ways Substances Cross Cell Membranes [p.56]

1. e, 2. l, 3. k, 4. b, 5. g, 6. c, 7. d, 8. i, 9. f, 10. a, 11. h, 12. j, 13. F, 14. T, 15. T, 16. F, 17. F, 18. T, 19. F, 20. F, 21. Isotonic solutions have the same solute concentration. A hypotonic solution has a lower solute concentration; a hypertonic solution has a higher solute concentration.

3.12. When Mitochondria Fail [p.57]

3.13. Metabolism: Doing Cellular Work [pp.58-59]

1. c, 2. d, 3. b, 4. g, 5. i, 6. h, 7. f, 8. j, 9. a, 10. e, 11. T, 12. F, 13. T, 14. F, 15. F, 16. T, 17. T, 18. F, 19. c, 20. a, 21. d, 22. b

3.14. How Cells Make ATP [pp.60-61]

3.15. Summary of Cellular Respiration [p.62]

3.16. Other Energy Sources [p.63]

3.17. No Thanks to Arsenic [p.63]

1. carbohydrates, 2. electrons, 3. aerobic, 4. glycolysis, 5. glucose, 6. cytoplasm, 7. two, 8. energy-requiring, 9. glucose, 10. PGAL, 11. four, 12. two, 13, does not,

14. a mitochondrion, 15. coenzyme A, 16. Krebs cycle, 17. CO_2, 18. two, 19. coenzyme, 20, electron transport system, 21. inner, 22. ATP, 23. water, 24. 36, 25. Arsenic disrupts the Krebs cycle and electron transport chain. ATP production stops and the cell dies, 26. Lactate fermentation requires no oxygen and produces only two molecules of ATP through glycolysis. Aerobic cellular respiration requires oxygen and produces 36 molecules of ATP., 27. Glycogen is the storage sugar found in humans. It is stored in muscle and liver cells., 28. Between meals or during exercise, the body may use triglycerides as energy alternatives to glucose. Fat is stored in adipose tissue, 29. Excess proteins are broken down to amino acids, 30. Muscle feels sore when lactic acid builds up in them.

Self-Quiz

1. d, 2. a, 3. d, 4. d, 5. d, 6. a, 7. a, 8. c, 9. b, 10. d, 11. c, 12. a, 13. c, 14. b, 15. d, 16. d, 17. c, 18. a

4 TISSUES, ORGANS, AND ORGAN SYSTEMS

Chapter Introduction [p.67]

4.1 Epithelium: The Body's Covering and Linings [pp. 68-69]

1. T, 2. F, 3. T, 4. T, 5. F 6. cells, 7. free, 8. basement, 9. simple, 10. stratified, 11. pseudostratified, 12. shape, 13. squamous, 14. cuboidal, 15. columnar, 16. gland, 17. exocrine, 18. endocrine, 19. hormones, 20. simple squamous, 21. simple cuboidal, 22. simple columnar

4.2 Connective Tissue: Binding, Support, and Other Roles [pp.70-71]

1. connective, 2. ground substance, 3. matrix, 4. collagen, 5. elastin, 6. cartilage, 7. bone, 8. adipose, 9. blood, 10. a, 11. a, 12. b, 13. c, 14. b, 15. a, 16. a, 17. a, 18. c, 19. a, 20. a, 21. b, 22. b, 23. a, 24. c, 25. a, 26. c, 27. b, 28. b, 29. loose connective tissue, 30. dense irregular connective tissue, 31. dense regular connective tissue, 32. cartilage, 33. bone tissue, 34. adipose tissue,

4.3. Muscle Tissue: Movement [p.72]

4.4. Nervous Tissue: Communication [p.73]

4.5. Healing with Stem Cells and Lab-Grown Tissues [p.73]

1. skeletal muscle, 2. cardiac muscle, 3. smooth muscle, 4. smooth, 5. cardiac, 6. skeletal, 7. smooth, 8. smooth, 9. cardiac, 10. skeletal, 11. skeletal, 12. d, 13. c, 14. a, 15. b, 16. f, 17. e, 18. stem cell, 19. replacement tissue, 20. biotechnology, 21. bone marrow, 22. umbilical cord blood

4.6. Cell Junctions: Holding Tissues Together [p.74]

4.7. Tissue Membranes: Thin, Sheetlike Covers [p.75]

1. tight junction, 2. adhering junction, 3. gap junction, 4. 2, 5. 1, 6. 3, 7. serous membrane, 8. synovial membrane, 9. cutaneous membrane, 10. mucous membrane, 11. mucous, 12. synovial, 13.

cutaneous, 14. mucous, 15. serous

4.8. Organs and Organ Systems [pp.76-77]
1. B, 2. C, 3. A, 4. D, 5. tissues, 6. cranial, 7. spinal, 8. thoracic, 9. abdominal, 10. pelvic, 11. organ systems, 12. k, 13. i, 14. c, 15. b, 16. a, 17. h, 18. d, 19. f, 20. j, 21. e, 22. g

4.9. The Skin: An Example of an Organ System [pp.78-79]
1. barrier to microbes, holds in moisture; repairs small wounds; helps regulate body temperature; has sensory receptors, makes the precursor to vitamin D, 2. Ultraviolet radiation stimulates melanocytes to produce melanin which gives some protection against UV radiation., 3. Vitiligo is a disorder in which melanocytes die and white patches form on the skin., 4. hair, 5. blood vessel, 6. pressure sensitive sensory receptor, 7. smooth muscle, 8. sweat gland, 9. hair follicle, 10. sebaceous gland, 11. epidermis, 12. dermis, 13. hypodermis, 14. c, 15. a, 16. b, 17. c, 18. b, 19. a, 20. F, 21. T, 22. F, 23. T, 24. T, 25. F, 26. B, 27. D, 28. E, 29. A, 30. C

4.10. Homeostasis: The Body in Balance [pp.80-81]
4.11. How Homeostatic Feedback Maintains the Body's Core Temperature [pp.82-83]
1. Extracellular, 2. interstitial, 3. metabolic waste products, 4. ions, 5. Homeostasis, 6. stability, 7. receptors, 8. stimulus, 9. integrator, 10. effectors, 11. "set points", 12. feedback, 13. negative, 14. positive, 15. stimulus, 16. sensory receptor, 17. integrator, 18. effector, 19. response, 20. b, 21. a, 22. b, 23. b, 24. a, 25. a, 26. E, 27. M, 28. L, 29. J, 30. D, 31. I, 32. F, 33. N, 34. K, 35. B, 36. H, 37. A, 38. C, 39. G

Self-Quiz
1. D, 2. G, 3. I, 4. J, 5. E, 6. L, 7. B, 8. H, 9. K, 10. M, 11. C, 12. F, 13. A, 14. c, 15. c, 16. d, 17. a, 18. b, 19. c, 20. d, 21. b, 22. b, 23. c, 24. d

5 THE SKELETAL SYSTEM
Chapter Introduction [p. 87]
5.1. Bone: Mineralized Connective Tissue [pp.88-89]
5.2. The Skeleton: The Body's Bony Framework [pp.90-91]
1a. periosteum, 1b. dense connective tissue, 1c. osteoblasts, 2. bone marrow, 3. spongy, 4. compact, 5. osteon, 6. blood vessels, 7a. osteocytes, 7b. lacunae, 8. living, 9. 3,1,5,2,4, 10. R, 11. D, 12. F, 13. O, 14. C, 15. G, 16. Q, 17. J, 18. E, 19, P, 20. N, 21. A, 22. M, 23. H, 24. B, 25. K, 26. I, 27. L, 28. osteoblasts, 29. osteoclasts, 30. matrix, 31. hard, 32. deposited, 33. remove, 34. osteoblasts, 35. calcium, 36. PTH, 37. calcitonin, 38. negative, 39. The diameter of a child's bones increases as osteoblasts from bone at the surface of each shaft. At the same time, osteoclasts break down a small amount of bone inside the shaft., 40. The two types of bone are spongy and compact. Compact bone is dense forms the shaft of long bones whereas spongy bone looks like a sponge and is found inside the shaft of long bones or at the ends., 41. yellow marrow, 42. irregular, 43. 206, 44. axial, 45. appendicular, 46. ligaments, 47. tendons, 48.

phosphorous, 49. movement, support, protection, mineral storage, blood cell formation, 50. skull bone, 51. sternum, 52. ribs, 53. vertebrae, 54. intervertebral disc, 55. clavicle, 56. scapula, 57. humerus, 58. radius, 59. ulna, 60. carpals, 61. metacarpals, 62. phalanges, 63. pelvic girdle, 64. femur, 65. patella, 66. tibia, 67. fibula, 68. tarsals, 69. metatarsals, 70. phalanges

5.3. The Axial Skelton [pp.92-93]
5.4. The Appendicular Skeleton [pp.94-95]
1. parietal bone, 2. temporal bone, 3. occipital bone, 4. frontal bone, 5. sphenoid bone, 6. zygomatic bone, 7. maxilla, 8. mandible, 9. hard palate, 10. foramen magnum, 11. cervical vertebrae, 12. thoracic vertebrae, 13. lumbar vertebrae, 14. sacrum, 15. coccyx, 16. axial, 17. eight, 18. sinuses, 19. Temporal, 20. sphenoid, 21. nose, 22. occipital, 23. occipital, 24. foramen magnum, 25. mandible, 26. zygomatic, 27. lacrimal, 28. maxillary, 29. vomer, 30. cervical, 31. true, 32. true, 33. true, 34. false, 35. true, 36. true, 37. false, 38. false, 39. false, 40 true

5.5. Joints: Connections Between Bones [pp.96-97]
5.6. Disorders of The Skeleton: [pp.98-99]
1. synovial, 2. synovial, 3. hingelike, 4. ball-and-socket, 5. cartilaginous, 6. fibrous, 7. abduction, 8. adduction, 9. circumduction, 10. rotation, 11. flexion, 12. extension, 13. supination, 14. pronation, 15. E, 16. I, 17. A, 18. F, 19. H, 20. D, 21. L, 22. J, 23.K, 24. G, 25. B, 26. M, 27. C

5.7. The Skeletal System in Homeostasis [p.100]
1. D, 2. H, 3. F, 4. B, 5. J, 6. A, 7. K, 8. E, 9. C, 10. I, 11. G

Self-Quiz
1. d, 2. g, 3. e, 4. j, 5. i, 6. c, 7. k, 8. h, 9. b, 10. i, 11. c, 12. e, 13. a, 14. f, 15. b, 16. d, 17. c, 18. a, 19. a, 20. b

6 THE MUSCULAR SYSTEM
Chapter Introduction [p. 103]
6.1 The Body's Three Kinds of Muscles [pp.104-105]
6.2 The Structure and Functioning of Skeletal Muscles [pp.106–107]
6.3. How Muscles Contract [pp.108–109]
1. attached to bones, 2. walls of hollow organs and tubes, 3. heart, 4.yes, 5. no, 6. yes, 7. yes, 8. no, 9. no, 10. These muscles are attached to the skeleton and move body parts., 11. triceps brachii (G), 12. pectoralis major (K), 13. serratus anterior (M), 14. external oblique (B), 15. rectus abdominis (F), 16. adductor longus (J), 17. sartorius (H), 18. quadriceps femoris (N), 19. tibialis anterior (C), 20. biceps brachii (E), 21. deltoid (D), 22. trapezius (L), 23. latissimus dorsi (O), 24. gluteus maximus (I), 25. biceps femoris (P), 26. gastrocnemius (A), 27. connective, 28. myofibrils, 29. origin, 30. insertion, 31. pulls, 32. fibers, 33. pairs, 34. contracts, 35. bend, 36. relax, 37. straighten, 38. synergistic, 39. wrist, 40. b, 41. a, 42. b, 43. a, 44. sarcomere, 45. skeletal muscle, 46. muscle cells, 47. myofibril, 48. striated, 49. Z bands, 50. thin filaments, 51. I band, 52. thick filaments, 53. H zone, 54. A band, 55. actin, 56.

myosin, 57. sliding filament, myosin, 58. actin, 59. ATP, 60. cross-bridge, 61. actin, 62. ATP, 63. hundreds of, 64. Stiffening of skeletal muscles after death, due to lack of ATP which prevents myosin cross-bridges from breaking apart from actin.

6.4. How the Nervous System Controls Muscle Contraction [pp.110–111]

6.5. How Muscle Cells Get Energy [p. 112]

6.6. Properties of Whole Muscles [pp.112–113]

6.7. Muscle Disorders [p.114-115]

6.8. Making the Most of Muscles [p.116]

6.9. Muscles and the Muscular System in Homeostasis [p.117]

1. F, 2. C, 3. A, 4. G, 5. E, 6. D, 7. B, 8. 5, 2, 4, 7, 9, 6, 8, 1, 3, 9. c, 10. a, 11. b, 12. a, 13. c, 14. c, 15. b, 16. I 17. J, 18. A, 19. E, 20. D, 21. B, 22. G, 23. C, 24. F, 25. H, 26. T, 27. F, 28. F, 29. T, 30. F, 31. F, 32. T, 33. T, 34. T, 35. cramps, 36. tics, 37. a disease of adults, 38. hot packs, 39. strain, 40. lengthening, 41. Both conditions affect skeletal muscles and cause their victims to have difficulty moving these muscles. Duchenne's generally affects children and causes their muscles to be unable to contract throughout their bodies. It is a fatal disease. Myotonic muscular dystrophy usually affects adults causing their muscles to contract strongly, but not relax. It generally affects only the hands and feet and is not life threatening., 42. T, 43. F, 44. T, 45. F, 46. F, 47. T, 48. C, 49, D., 50. F, 51. A, 52. E, 53. B

Self-Quiz

1. d, 2. e, 3. a, 4. b, 5. c, 6. b, 7. d, 8. a, 9. e, 10. c

---

7 CIRCULATION: THE HEART AND BLOOD VESSELS

Chapter Introduction [p. 121]

7.1. The Cardiovascular System: Moving Blood Through The Body [pp. 122-123]

7.2. The Heart: A Muscular Double Pump [pp. 124-125]

7.3. The Two Circuits of Blood Flow [pp. 126-127]

7.4. How Cardiac Muscle Contracts [p. 128]

1. cardiovascular; 2. heart; 3. blood vessels; 4. arteries; 5. arterioles; 6. capillaries; 7. venules; 8. veins; 9. rapidly; 10. slowly; 11. oxygen; 12. nutrients; 13. wastes; 14. Homeostasis; 15. extracellular; 16. lymphatic; 17. jugular; 18. superior; 19. pulmonary; 20. hepatic portal; 21. renal; 22. inferior; 23. iliac; 24. femoral; 25. femoral; 26. iliac; 27. abdominal; 28. renal; 29. brachial; 30. coronary; 31. pulmonary; 32. ascending; 33. carotid; 34. D; 35. K; 36. G; 37. E; 38. J; 39. F; 40. H; 41. B; 42. I; 43. C; 44. L; 45. M; 46. A; 47. N; 48. superior vena cava; 49. pulmonary valve; 50. right atrioventricular (or tricuspid) valve; 51. inferior vena cava; 52. aorta; 53. left ventricle; 54. endocardium; 55. Myocardium; 56. pericardium; 57. 1, 5, 2, 7, 8, 3, 6, 4; 58. D; 59. F; 60. C; 61. J; 62. A; 63. H; 64. E; 65. G; 66. B; 67. 1; 68. D; 69. pulmonary; 70. systemic; 71. arteries; 72. organs; 73. arteries; 74. arterioles; 75. capillaries; 76. venules; 77. veins; 78. superior vena cava; 79. inferior vena cava; 80.

capillary beds; 81. hepatic portal vein; 82. impurities; 83. hepatic; 84. oxygenated; 85. intercalated disks; 86. cardiac conduction; 87. sinoatrial; 88. atrioventricular; 89. Purkinje fibers; 90. SA; 91. nervous.

7.5. Blood Pressure [p. 129]

7.6. Structure and Functions of Blood Vessels [pp. 130-131]

7.7. Capillaries: Where Substances Move between Blood and Tissues [pp. 132-133]

7.8. Cardiovascular Diseases and Disorders [pp.134-135]

7.9. Infections, Cancer, and Heart Defects [p. 136]

7.10. The Cardiovascular System and Blood in Homeostasis [p.137]

1. blood pressure; 2. aorta; 3. systemics; 4. systolic; 5. aorta; 6. diastolic; 7. aorta; 8. hypertension; 9. hypotension; 10. circulatory shock; 11. a; 12. c; 13. b; 14. c; 15. a; 16. b; 17. c; 18. e; 19. d; 20. e; 21. a; 22. e; 23. e; 24. d; 25. artery, elastic tissue; 26. arteriole, smooth muscle; 27. capillary, endothelium; 28. venule, smooth muscle; 29. vein, valve; 30. D; 31. C; 32. A; 33. B; 34. greater; 35. T; 36. T; 37. lymphatic; 38. from arterioles into; 39. T; 40. inflammation; 41. C-reactive protein; 42. homocysteine; 43. Arteriosclerosis; 44. Atherosclerosis; 45. atherosclerotic; 46. embolus; 47. HDLs; 48. LDLs; 49. coronary artery; 50. Laser; 51. Balloon; 52. aneurysm; 53. attack; 54. attack; 55. ECG; 56. arrhythmias; 57. bradycardia; 58. Tachycardia; 59. fibrillation; 60. watching intake of cholesterol and trans fat, getting regular exercise, and not smoking; 61. rheumatic fever; 62. heart valves; 63. autoimmune; 64. heart valves; 65. endocarditis; 66. Lyme disease; 67. "bull's-eye" rash; 68. heart muscle; 69. myocarditis; 70. heart muscle; 71. Cancer; 72. heart; 73. vessels; 74. "blue babies," 75. The cardiovascular system distributes blood throughout the body allowing blood to distribute needed materials to and removing unneeded materials from cells.

Self-Quiz

1. d; 2. c; 3. c; 4. b; 5. a; 6. d; 7. a; 8. e; 9.c; 10. d

---

8 BLOOD

Chapter Introduction [p.141]

8.1. Blood: Plasma, Blood Cells and Platelets [p.142-143]

1. 4-5, 2. plasma, 3. water, 4. albumin, 5. water, 6. proteins, 7. pH, 8. volume, 9. red blood, 10. hemoglobin, 11. stem cells, 12. white blood, 13. foreign, 14. disease agents, 15. tissues, 16. bone marrow, 17. granulocytes, 18. neutrophils, 19. eosinophils, 20. basophils, 21. agranulocytes, 22. monocytes, 23. lymphocytes, 24. megakaryocytes, 25. platelets, 26. clotting, 27. leukocytes, 28. platelets, 29, neutrophil, 30. 8%, 31. lymphocytes, 32. c, 33. b, 34. b

8.2. How Blood Transports Oxygen [p.144]

8.3 Making New Red Blood Cells [p.145]

1. hemoglobin, 2. oxyhemoglobin, 3. more oxygen; cooler; less acidic, 4. lungs, 5. less oxygen; warmer; more acidic, 6. tissues, 7. hemoglobin, 8. red blood cells (erythrocytes), 9. four, 10. iron, 11. carries oxygen, 12. oxyhemoglobin, 13. f, 14. g, 15. a, 16. b, 17. c, 18. d, 19. h, 20. e

8.4. Different Red Blood Cells [pp.146-147]

8.5. Rh Blood Typing [p.148]

8.6. What Can Your Blood Say About You? [p.149]

1. i, 2. j, 3. b, 4. a, 5. f, 6. g, 7. h, 8. c, 9. d, 10. e, 11. A, 12. B, 13. A and B, 14. neither A nor B, 15. Anti-B, 16. Anti-A, 17. neither Anti-A nor Anti-B, 18. both Anti-A and Anti-B, 19. Rh⁻, 20. the person receiving the transfusion, 21. type AB; 22. b; 23. d; 24. e; 25. a; 26. c

8.7. Hemostasis and Blood Clotting [pp.150-151]

8.8. Blood Disorders [pp.152-153]

1. b, 2. e, 3. c, 4. d, 5. a, 6. f, 7. b, 8. a, 9. f, 10. g, 11. d, 12. h, 13. c, 14. e, 15. d, 16. f, 17. j, 18. k, 19. b, 20. e, 21. i, 22. c, 23. l, 24. a, 25. h, 26. g

Self-Quiz

1. d, 2. e, 3. a, 4. e, 5. e, 6. a, 7. b, 8. d, 9. a, 10. b, 11. c, 12. d, 13. a, 14. d, 15. e

9 IMMUNITY

Chapter Introduction [p.155]

9.1. Overview of Body Defenses [pp.156-157]

9.2. The Lymphatic System [pp.158-159]

1. viruses, bacteria, fungi, protozoa, parasitic worms; 2. Pathogens have antigens on their surfaces that cause an immune response; 3. immunity; 4. a; 5. c; 6. b; 7. b; 8. b; 9. a; 10. c; 11. c; 12. b; 13, c; 14. cytokines; 15. basophils; 16. mast cells; 17. antibodies; 18. females; 19. males; 20. autoimmune; 21. multiple sclerosis, lupus erythematosus, Hashimoto's thyroiditis; 22. D; 23. G; 24. B; 25. E; 26. F; 27. C; 28. A; 29. tonsils; 30. right lymphatic duct; 31. thymus; 32. thoracic duct; 33. spleen; 34. lymph vessels; 35. lymph nodes; 36. Tissue fluid is fluid lost from capillaries. Once it is moved into lymph vessels, the fluid is called lymph; 37. c; 38. a; 39. b; 40. a; 41. b; 42. c; 43. 2, 1, 4, 3; 44. b; 45. c; 46. a; 47. b; 48. a; 49. b; 50. tonsils, patches of tissue in small intestine, appendix, airways leading to lungs.

9.3. Barriers to Infection [p.160]

9.4. Innate Immunity [pp.160-161]

9.5. Overview of Adaptive Defenses [pp.162-163]

1. They grow so densely that they prevent harmful species from gaining a foothold, but if they are reduced in number, the harmful pathogens can cause an infection; 2. A; 3. D; 4. C; 5. B; 6. innate; 7. macrophages; 8. anything; 9. cytokines; 10. macrophages; 11. complement; 12. pathogens; 13, phagocytes; 14. "complement coat"; 15. phagocyte; 16. membrane attack complexes; 17. cytokines; 18. Mast cells; 19. more; 20. edema; 21. interleukins, prostaglandins; 22. It makes the body too hot for many pathogens to function normally; 23. D; 24. A; 25. C; 26. B; 27. third line of defense which recognizes

self vs nonself, is specific, diverse and has memory; 28. bone marrow; 29. thymus gland; 30. thymus gland; 31. lymphoid tissue; 32. thymus, spleen, lymphatic tissue; 33. thymus, spleen, lymphatic tissue; 34. antibody-mediated; 35. cell-mediated; 36. T; 37. T cells; 38. T; 39. T; 40. engulfs; 41. MHC markers; 42. T; 43. T; 44. T

9.6. Antibody-Mediated Immunity: Defending Against Threats Outside Cells [pp.164-165]

9.7. Cell-Mediated Responses: Combating Threats Inside Cells [pp.166-167]

1. antibodies; 2. Y; 3. antigen; 4. genes; 5. plasma membrane; 6. arms; 7. complement; 8. lymph node; 9. endocytose (engulf); 10. Antigen-MHC complexes; 11. helper; 12. activate; 13. divide; 14. plasma; 15. phagocytes; 16. memory; 17. five; 18. immunoglobulins; 19. E; 20. D; 21. B; 22. A; 23. C; 24. cell; 25. T; 26. T; 27. T; 28. do; 29 cytotoxic; 30. T; 31. will not; 32. MHC markers on donor cells may be similar to the recipient's but are different enough to cause rejection due to immune response.

9.8. Applications of Immunology [pp.168-169]

9.9. Immune System Disorders [pp.170-171]

9.10. HIV and AIDS [pp.172-173]

9.11. Patterns of Infectious Disease [pp.174-175]

1. G; 2. H; 3. A; 4. E; 5. I; 6. F; 7. B; 8. C; 9. D; 10. J; 11. allergy; 12. allergen; 13, predisposed; 14. antibodies; 15. mast; 16. histamine; 17. constrict; 18. Hay fever; 19. invader; 20. anaphylactic shock; 21. cardiovascular; 22. epinephrine; 23. Antihistamines; 24. IgG; 25. autoimmune; 26. Rheumatoid arthritis; 27. diabetes; 28. systemic lupus erythematosus; 29. lymphocytes; 30. severe combined immune deficiency; 31. human immunodeficiency virus; 32. acquired immunosuppressive syndrome; 33. body fluids; 34. macrophages; 35. helper T; 36. dendritic cells; 37. cancer; 38. no; 39. severely depressed; 40. T; 41. T; 42. T; 43. not effectively; 44. RNA; 45. proteins; 46. no; 47. rapidly; 48. T; 49. T; 50; direct contact, indirect contact, inhaling pathogens, contact with a vector; 51. infection acquired in a hospital, those hospitalized are likely to have compromised immune systems, invasive procedures give bacteria easy access to tissues, antibiotic-resistant pathogens due to intensive antibiotic use; 52. D; 53. E; 54. A; 55. B; 56. C.

Self-Quiz

1. d; 2. c; 3. a; 4. d; 5. a; 6. e; 7. e; 8. c; 9. a; 10. b; 11. H; 12. D; 13. E; 14. J; 15. B; 16. G; 17. C; 18. I; 19. F; 20. A.

10 THE RESPIRATORY SYSTEM

Chapter Introduction [p.179]

10.1. The Respiratory System – Built for Gas Exchange [pp.180-181]

1. toxic; 2. lung cancer; 3. 70; 4. doubles; 5. blood pressure; 6. LDL cholesterol; 7. HDL cholesterol; 8. oral cavity (H); 9. pleural membrane (J); 10. intercostal muscles; (F); 11. diaphragm (B); 12. nasal

cavity (E); 13. pharynx (K); 14. epiglottis (D); 15. larynx (A); 16. trachea (G); 17. lung (L); 18. bronchial tree (C); 19. bronchiole; 20. alveolar duct; 21. alveoli, (I); 22. alveolar sac; 23. pulmonary capillaries; 24. B; 25. H; 26. E; 27. K; 28. J; 29. I; 30. N; 31. C; 32. O; 33. M; 34. F; 35. A; 36. D; 37. G; 38. L; 39. D; 40. E; 41. B; 42. G; 43. C; 44. F; 45. A.

10.2. Respiration = Gas Exchange [pp.182-183]
10.3. Breathing at High Altitude and Underwater [p.183]
10.4. Breathing: Air In, Air Out [184-185]
1. Gas exchange; 2. pressure; 3. 21; 4. carbon dioxide; 5. respiratory; 6. dissolved; 7. pressure gradient; 8. alveoli; 9. hemoglobin; 10. four; 11. releases; 12. gradient; 13. decrease; 14. hyperventilation; 15. increase; 16. left; 17. right; 18 intercostals; 19. diaphragm; 20. down; 21. upward and outward; 22. expands; 23. up; 24. inward and downward; 25. expiration; 26. chest cavity; 27. enters; 28. expiration; 29. out of; 30. chest; 31. pleural sac; 32. pleural; 33. tidal volume; 34. inspiratory; 35. expiratory; 36. vital capacity; 37. one-half; 38. residual; 39. 350; 40. trachea

10.5. How Gases are Exchanged and Transported [pp.186-187]
10.6. Controls over Breathing [pp.188-189]
10.7. Respiratory System Disorders: Tobacco, Irritants and Apnea [pp.190-191]
10.8. Pathogens and Cancer in the Respiratory System [p.192]
10.9. The Respiratory System in Homeostasis [p.193]
1. H; 2. J; 3. B; 4. D; 5. K; 6. G; 7. C; 8. A; 9. E; 10. I; 11. L; 12. F; 13. 70; 14. more; 15. decreases; 16. internal; 17. bicarbonate; 18. water; 19. falls; 20. lower; 21. buffer; 22. oxygen; 23. carbon dioxide; 24. arteries; 25. diaphragm; 26. medulla; 27. cerebrospinal; 28. respiratory; 29. CO2 ; 30. carotid; 31. aortic; 32. ventilation; 33. increase; 34. relax; 35. constrict; 36. involuntary; 37. secondhand; 38. 80; 39. Each individual will have his or her own responses; 40. F; 41. E; 42. G; 43. A; 44. D; 45. B; 46. C; 47. Delivers oxygen to cells for cellular respiration and eliminates waste carbon dioxide produced by as cells make ATP.

Self-Quiz
1. c; 2. e; 3. a; 4. b; 5. a; 6. a; 7. a; 8. a; 9. c; 10. e

---

11 DIGESTION AND NUTRITION
Chapter Introduction [p.197]
11.1. Overview of the Digestive System [p.198-199]
11.2. Chewing and Swallowing: Food Processing Begins [pp.200-201]
11.3. The Stomach: Food Storage, Digestion, and More [p.202]
1. pharynx; 2. esophagus; 3. liver; 4. gall bladder; 5. pancreas; 6. large intestine (colon); 7. rectum; 8. anus; 9. mouth (oral cavity); 10. salivary glands; 11. stomach; 12. small intestine; 13. a. mouth cavity; b. salivary glands; c. pharynx; d. esophagus; e. stomach;

f. small intestine; g. pancreas; h. liver; i. gallbladder; j. large intestine; k. rectum; l. anus; 14. lumen (C); 15. mucosa (E); 16. submucosa (A); 17. smooth muscle layer (D); 18. serosa (B); 19. Circular arrays of smooth muscle function to close off a passageway, e.g., to prevent backflow; 20. salivary glands, liver, gall bladder, pancreas; 21. C; 22. I; 23. B; 24. E; 25. H; 26. J; 27. A; 28. D; 29. G; 30. F; 31. pharynx; 32. involuntary; 33. epiglottis; 34. sphincter; 35. stomach; 36. small intestines; 37. hydrochloric acid; 38. pepsins; 39. intrinsic; 40. gastric juice; 41. chyme; 42. proteins; 43. denatures; 44. pepsins; 45. gastrin; 46. gastric mucosal barrier; 47. peristalsis; 48. pyloric; 49. close; 50. small intestine; 51. rugae; 52. water; 53. alcohol.

11.4. The Small Intestine: A Huge Surface for Digestion and Absorption [p.203]

11.5. Accessory Organs: The Pancreas, Gall Bladder, and Liver [pp.204-205]
11.6. Digestion and Absorption in the Small Intestine [pp.206-207]
11.7. The Large Intestine [p.208]
1. wall; 2. surface area; 3. villi; 4. blood vessels; 5. bloodstream; 6. microvilli; 7. brush border; 8. Gland cells; 9. b; 10. c; 11. a; 12. c; 13. a; 14. b; 15. a; 16. b; 17. c; 18. hepatic portal vein; 19. hepatic vein; 20. glycogen; 21. toxins; 22. urea; 23. a, b, c; 24. b, c; 25. c; 26. c; 27. d; 28. a; 29. d; 30. c, d; 31. d; 32. b; 33. c, d; 34. Fats do not dissolve in water, so bile salts coat fat droplets and keep them separated so that there is more surface area on which digestive enzymes can work; 35. It creates a back and forth movement that mixes digested material and forces it against the small intestine wall; 36. small intestine; 37. osmosis; 38. Transport proteins; 39. glucose; 40. amino acids; 41. bile salts; 42. micelles; 43. lacteals; 44. G; 45. F; 46. E; 47. H; 48. C; 49. D; 50. B; 51. A

11.8. Controls Over Digestion [p. 209]
11.9. Digestive System Disorders [p.210-211]
11.10. Infections in the Digestive System [p.212]
11.11. The Digestive System in Homeostasis [p.213]
1. D; 2. E; 3. B; 4. A; 5. C; 6. B; 7. A; 8. F; 9. G; 10. Q; 11. J; 12. P; 13. E; 14. K; 15. L; 16. O; 17. S; 18. C; 19. I; 20. N; 21. H; 22. D; 23. M; 24. R; 25. D; 26. E; 27. A; 28. F; 29. B; 30. C; 31. The digestive system breaks down bulk food to nutrients, vitamins, and minerals that can be absorbed into the bloodstream and delivered to cells for energy and building new cells and cell parts. It also absorbs water and stores and eliminates solid wastes as feces.

11.12. The Body's Nutritional Requirements [pp.214-215]
11.13. Vitamins and Minerals [pp.216-217]
11.14. Food Energy and Body Weight [pp.218-219]
11.15. Dealing with Weight Extremes [pp.220-221]
1. b; 2. a; 3. c; 4. a; 5. c; 6. b; 7. a; 8. b; 9. c; 10. b; 11. They add to caloric intake but fill no other nutritional needs; 12. Within minutes of absorption, there is a surge in blood levels of sugar and insulin; 13.

Essential means that our cells cannot make them, so they must be taken in with our diets; 14. We should decrease the intake of refined grains, trans fats and saturated fats, and refined sugars. We should increase intake of whole grains, legumes, dark green and orange vegetables, fruits, and milk products; 15. organic, inorganic; 16. any; 17. do; 18. inhibiting; 19. is; 20. fat-; 21. B, 22. C; 23. coenzymes; 24. C; 25. B; 26. A; 27. D; 28. 2100, 2700, 1100; 29. 25-year-old female; 30.37 year old male; 31. 59-year-old female; 32. about 6 hours; 33. type 2 diabetes, heart disease, osteoarthritis, high blood pressure, kidney stones; 34. Anorexia nervosa involves a person starving and over-exercising based on the perception that he or she is fat. Bulimia involves binging on huge amounts of food and then purging by vomiting or using laxatives.

Self-Quiz

1. e; 2. d; 3. b; 4. b; 5. a; 6. e; 7. b; 8. c; 9. b; 10. b; 11. d; 12. d; 13. b; 14. E; 15. D; 16. A; 17. G; 18. B; 19. C; 20. F

---

12 THE URINARY SYSTEM

Chapter Introduction[p.225]

12.1. The Challenge: Shifts in Extracellular Fluid [pp.226-227]

12.2. The Urinary System: Built for Filtering and Waste Disposal [pp.228-229]

12.3. How Urine Forms: Filtration, Reabsorption, and Secretion [pp.230-231]

1. The urinary system maintains stable conditions in the extracellular fluid by removing substances that enter the ECF from the intracellular fluid and adding needed substances that move from the ECF into the intracellular fluid; 2. a. absorption from liquids and solid food; b. metabolic reactions; c. excretion in urine; d. evaporation from lungs and skin; e. sweating; f. elimination in feces; g. absorption from liquids and solid food; h. secretion from cells; i. respiration; j. metabolism; k. urinary excretion; 1. respiration; m. sweating; 3. brain; 4. urinary excretion; 5. T; 6. respiratory; 7. T; 8. urine; 9. T; 10. protein; 11. urea; 12. T; 13. kidney; 14. ureter; 15. urinary bladder; 16. urethra; 17. renal cortex; 19. renal medulla; 19. ureter; 20. produce erythropoietin, convert vitamin D to a form that causes absorption of calcium, make rennin, remove metabolic wastes from blood and adjust fluid balance; 21. nephrons; 22. solutes; 23. blood; 24. epithelial; 25. active transport; 26. glomerulus; 27. Bowman's; 28. proximal; 29. Henle; 30. distal; 31. renal arteries; 32. capillaries; 33. permeable; 34. efferent; 35. peritubular; 36. venules; 37. cortex; 38. medulla; 39. glomerular capillaries; 40. proximal tubule; 41. distal tubule; 42. peritubular capillaries; 43. collecting duct; 44. Bowman's capsule.

12.4. How Kidneys Help Manage Fluid Balance and Blood Pressure [pp.232-233]

12.5. Removing Excess Acids and Other Substances in Urine [pp.234]

12.6. Kidney Disorders: When Kidneys Fail [p.235]

12.7. Cancer, Infections, and Drugs in the Urinary System [p.236]

12.8. The Urinary System in Homeostasis [p.237]

1. a. glomerulus; b. water, solutes; c. Bowman's, proximal; d. proximal; e. water; f. peritubular; g. nephron; h. capillaries; i. urine; 2. high; 3. large; 4. tissue fluid; 5. osmosis; 6. Secretion; 7. antibiotics; 8. urinalysis; 9. water; 10. urinary bladder; 11. relaxes; 12. external; 13. 2/3; 14. medulla; 15. out of; 16. T; 17. ascending; 18. salt; 19. inner; 20. T; 21. brain; 22. distal; 23. more; 24. loss; 25. saliva; 26. 4,2,3,1; 27. 7.37 to 7.43; 28. excrete; 29. bicarbonate; 30. blood; 31. nephron tubule; 32. peritubular capillaries; 33. H+; 34. basic; 35. less; 36. ammonia; 37. excreted; 38. E; 39. G; 40. B; 41. C; 42. A; 43. D; 44. F; 45. The female urethra is much shorter than that of a male and its outer opening is close to the anus, making it easier for bacteria to get to the bladder; 46. painkillers; 47. alcohol; 48. illegal drugs; 49. Urinalysis; 50. diabetes; 51. urinary tract; 52. The kidneys adjust the chemical composition of extracellular fluids essential for the body and remove nitrogen wastes, maintain water, electrolyte and acid and base balance. The bladder, ureters, and urethra provide for the storage and elimination of wastes in urine.

Self-Quiz

1. d; 2. c; 3. e; 4. b; 5. d; 6. d; 7. d; 8. c; 9. e; 10. a

---

13 THE NERVOUS SYSTEM

Chapter Introduction [p. 241]

13.1. Neurons: The Communication Specialists [p.242-243]

13.2. Nerve Impulses = Action Potentials [pp.244-245]

13.3. How Neurons Communicate [pp.246-247]

1. E, 2. I, 3. B, 4. D, 5. C, 6. G, 7. A, 8. F, 9. H, 10. J, 11. Neuroglia are support cells in both the central and peripheral nervous system. They maintain ion concentrations, support and protect neurons, provide insulation to neurons and "clean-up" the central nervous system., 12. dendrites, 13. cell body, 14. trigger zone, 15. axon, 16. axon terminals, 17. channel proteins, 18. concentrations, 19. closed, 20. potassium, 21. in, out, 22. negatively, 23. potential, 24. reverse, 25. open, 26. positive, 27. action potential, 28. trigger, 29. gland, 30. T, 31. away from, 32. T, 33. active transport, 34. into, 35. T, 36. potassium, 37. resting level, 38. threshold level, 39. action potential, 40. neurotransmitters, 41. chemical synapse, 42. presynaptic, 43. calcium, 44. vesicles, 45. postsynaptic, 46. action potential, 47. inhibitory, 48. acetylcholine, 49. neurotransmitter, 50. endorphins, 51. pain, 52. endorphins 53. "high", 54. input zone, 55. graded potentials, 56. depolarize, 57. hyperpolarize, 58. summation, 59. Integration, 60. removal, 61. enzymes, 62. pumped.

13.4. Information Pathways [pp.248–249]

13.5. Overview of The Nervous System [pp.250–251]
1. nerve, 2. myelin sheath, 3. action potential, 4. Schwann, 5. node, 6. saltatory, 7. propagate, 8. central nervous system, 9. oligodendrocytes, 10. reflexes, 11. stereotyped, 12. reflex arcs, 13. contracts, 14. interneurons, 15. spinal cord, 16. interneurons, 17. motor, 18. circuits, 19. diverge, 20. reverberating, 21. axon, 22. myelin sheath, 23. neuron, 24. blood vessels, 25. nerve, 26. F, 27. E, 28. G, 29. D, 30. B, 31. C, 32. A, 33. central nervous system, 34. peripheral nervous system, 35. somatic, 36. autonomic, 37. spinal, 38. cranial, 39. ganglia, 40. cranial, 41. spinal cord, 42. cervical, 43. thoracic, 44. lumbar, 45. sacral, 46. coccygeal, 47. I, 48. II, 49. VIII, 50. V, 51. VII.

13.6. Major Expressways: Peripheral Nerves and the Spinal Cord [pp.252–253]

13.7. The Brain: Command Central [pp.254–255]

13.8. A Closer Look at the Cerebrum [pp.256–257]
1. A, 2. S, 3. S, 4. A, 5. A, 6. S, 7. A, 8. Parasympathetic, 9. Sympathetic, 10. Parasympathetic, 11. Sympathetic, 12. Parasympathetic, 13. gray matter, 14. gray matter, 15. meninges, 16. autonomic reflexes, 17. the brain, 18. spinal cord, 19. ganglion, 20. spinal nerve, 21. vertebra, 22. intervertebral disc, 23. meninges, 24. gray matter, 25. white matter, 26. brain, 27. cranium, 28. meninges, 29. dura, 30. hemispheres, 31. arachnoid, 32. pia, 33. cushions, 34. cerebrum, 35.corpus callosum, 36. F,B, 37. F, A, 38. H, E, 39. H,F, 40. H,D, 41. F, C, 42. cerebrospinal, 43. ventricles, 44. spinal cord, 45. blood-brain barrier, 46. lipid, 47. cerebrum, 48. T, 49. right, 50. T, 51. nerve tracts, 52. T, 53. cerebral cortex, 54. T, 55. learned, 56. T, 57. parietal, 58. T, 59. T, 60. inside, 61. T, 62. pia mater, 63. T, 64. the primary motor cortex, 65. thalamus, 66. hypothalamus, 67. corpus callosum, 68. midbrain, 69. pons, 70. medulla, 71. cerebellum.

13.9. Consciousness [p.258]

13.10. Memory [p.259]

13.11. Disorders of the Nervous System [pP.260-261]

13.12. The Brain on "Mind-Altering" Drugs [p.262]

13.13. The Nervous System in Homeostasis [P.263]
1. C, 2. D, 3. H, 4. E, 5. I, 6. B, 7. A, 8. F, 9. G, 10. C, 11. E, 12. G, 13. A, 14. F, 15. D, 16. B, 17. An addict must increase intake of the drug to stay ahead of the liver's activity., 18. Continued drug use over a period of time that maintains the self-perception of functioning normally., 19. B, 20. E, 21. C, 22. A, 23. D
Self-Quiz
1. a, 2. d, 3. a, 4. b, 5. a, 6. d, 7. e, 8. b, 9. d, 10. c.

---

14 SENSORY SYSTEMS
Chapter Introduction [p.267]

14.1 Sensory Receptors and Pathways [pp.268-269]

14.2 Somatic Sensations [pp.270-271]
1. A sensation is conscious awareness of a stimulus. A perception is understanding what the sensation means., 2. The brain's assessment of a given stimulus depends on (a) which sensory area receives signals from nerves, (b) the frequency of signals, and (c) the number of axons that respond to the stimulus., 3. Sensory adaptation occurs with the pressure sensations, such as the pressure from clothing on the skin., 4. Somatic senses involve receptors that are found at more than one location in the body, such as skin and skeletal muscles. Special senses involve receptors restricted to sense organs, such as the eyes or ears., 5. a, 6. d, 7. c, 8. a, 9. b, 10. a, 11. e, 12. f, 13. b, 14. d, 15. b, 16. f, 17. b, 18. cerebrum, 19. greatest, 20. mechanoreceptors, 21. Meissner's corpuscles, 22. Ruffini endings, 23. visceral pain, 24. activate pain receptors, 25. referred pain

14.3 Taste and Smell: Chemical Senses [pp.272-273]

14.4 Tasty Science [p.273]
1. chemoreceptors, 2. thalamus, 3. taste buds, 4. five, 5. salty, 6. umami, 7. olfactory, 8. olfactory bulbs, 9. cerebral cortex, 10. vomeronasal, 11. pheromones, 12. tastant, 13. two, 14. sensitive

14.5 Hearing: Detecting Sound Waves [pp.274-275]

14.6 Balance: Sensing the Body's Natural Position [pp.276-277]

14.7 Disorders of the Ear [p.277]
1. air, 2. mechanical, 3. amplitude, 4. frequency, 5. mechanoreceptors, 6. auditory canal, 7. membrane, 8. hairs, 9. action potential, 10. sound, 11. b, 12. c, 13. a, 14. c, 15. a, 16. b, 17. i, 18. h, 19. d, 20, g, 21. b, 22. f, 23. j, 24. c, 25. e, 26. a, 27. middle ear bones, 28, auditory canal, 29. tympanic membrane (eardrum), 30. round window, 31. cochlea, 32. auditory nerve, 33. oval window, 34.I, 35. H, 36.C, 37.E, 38.D, 39.B, 40.F, 41.A, 42. G

14.8 Vision: An Overview [pp.278-279]

14.9 From Visual Signals to "Sight" [pp.280-281]

14.10 Disorders of the Eye [pp.282-283]
1. photoreceptors, 2. visual cortex, 3. vision, 4. eyes, 5. cornea, 6. iris, 7. pupil, 8. lens, 9. retina, 10. optic nerve, 11. H, 12.K, 13.A, 14.B, 15.E, 16.G, 17.J, 18.F, 19.D, 20. I, 21.C, 22. choroid, 23. vitreous humor, 24. ciliary muscle, 25. iris, 26. pupil, 27. lens, 28. cornea, 29. aqueous humor, 30. sclera, 31. retina, 32. fovea, 33. optic disk (blind spot), 34. optic nerve, 35. cones, 36. rods/cones, 37. vitamin A, 38. fovea, 39. ganglion cells, 40. retinas, 41. visual field, 42. criss-cross, 43.N, 44.I, 45.D, 46.F, 47.H, 48.G, 49.C, 50.E, 51.A, 52.L, 53. J, 54.K, 55.M,
Self-Quiz
1. a, 2. c, 3. c, 4. a, 5. d, 6. c, 7. d, 8. a, 9. b, 10. d

---

15 THE ENDOCRINE SYSTEM
Chapter Introduction [p.287]

15.1. The Endocrine System: Hormones [pp.288-289]

15.2. Types of Hormones and Their Signals [pp.290-291]
1. E, 2. D, 3. G, 4. F, 5. A, 6. B, 7. C, 8. a. (9) PTH, b. (4) cortisol, aldosterone, sex hormones c. (4) epinephrine, norepinephrine, d. (11) insulin, glucagon, e. (5) estrogen, progesterone, f. (6) testosterone, g. (1) six releasing and inhibiting

hormones, ADH, oxytocin, h. (2) ACTH, TSH, FSH, LH, prolactin, growth hormone, i. (3) antidiuretic hormone, oxytocin, j. (8) thyroxine, triiodothyronine, k. (10) thymosins, l. (7) melatonin, 9. They bind to protein receptors of target cells, 10. a, 11. b, 12. a, 13. a, 14. b, 15. b, 16. a, 17. b

15.3. The Hypothalamus and Pituitary Gland [pp.292-293]

15.4. Growth Hormone Functions and Disorders [p.294]

15.5. Hormones as Long-Term Controllers [p.295]
1. A (G), 2. P (F), 3. A (A), 4. A (H), 5. A (C), 6. P (D), 7. A (B), 8. A. (E), 9. hypothalamus, 10. posterior, 11. anterior, 12. releaser, 13. inhibitor, 14. oxytocin, 15. GH, 16. Pituitary dwarfism, 17. gigantism, 18. acromegaly

15.6. The Thyroid and Parathyroid Glands [pp.296-297]

15.7. Adrenal Glands and Stress Responses [pp.298-299]

15.8. The Pancreas: Regulating Blood Sugar [p.300]

15.9. Blood Sugar Disorders [p.301]
1. C, 2. E, 3. D, 4. G, 5. H, 6. A, 7. B, 8. F, 9. A, 10. C, 11. B, 12. If a person is deficient in iodine, low levels of thyroid hormones cause secretion of TSH. The thyroid attempts to make hormones but can't causing continued release of TSH and thyroid stimulation. The gland ends up becoming enlarged., 13. Grave's disease is caused by hyperthyroidism. This may occur due to a tumor in or inflammation of the thyroid gland, or an autoimmune disorder where antibodies stimulate thyroid cells., 14. Cortisol promotes the breakdown of muscle proteins and stimulates the liver to take up amino acids and synthesize glucose., 15. Hypoglycemia may be caused by anything that raises blood insulin levels such as miscalculated insulin injections or an insulin-secreting tumor. This may cause insulin shock where the brain "stalls" due to lack of fuel.., 16. thyroid, 17. metabolic, 18. calcitonin, 19. iodine, 20. TSH, 21. goiter, 22. hypothyroidism, 23 overweight., 24. iodized salt, 25. Grave's., 26. toxic, 27. hyperthyroidism, 28. increased, 29. parathyroid, 30. parathyroid, 31. calcium, 32. remodeling, 33. nephrons, 34. D, 35. rickets, 36. bones, 37. kidney stones, 38. muscles, 39. E, 40. H, 41. F, 42. A, 43. B, 44. G, 45. D, 46. I, 47. C, 48. Causes suppression of the immune system by release of cortisol. The individual becomes more susceptible to disease.., 49. exocrine, endocrine, 50. islets, 51. alpha, 52. raise, 53. beta, 54. lower, 55. inhibit, 56. rises, 57. excessive, 58. energy, 59. decreased, 60. a viral infection, 61. type 1 diabetes, 62. type 1 diabetes, 63. type 2 diabetes, 64. obesity, 65. capillaries, 66. death, 67. is at risk for, 68. can

15.10 Other Hormone Sources [p.302]

15.11 The Endocrine System in Homeostasis [p.303]
1. Testosterone contributes to libido in the female. It is produced in the ovaries., 2. Estrogen and progesterone are involved in the proper development of sperm. They are produced by the testes., 3. Melatonin influences the biological clock. The longer the days and shorter the nights, the less melatonin is produced and vice versa., 4. When blood pressure rises, the atria secrete ANP which causes the kidneys to reabsorb less sodium and water. More water ends up in the urine, this lowering the pressure., 5. C, 6. A, 7. F, 8. E, 9. B, 10. D

Self-Quiz
1. a, 2. e, 3. d, 4. a, 5. d, 6. c, 7. e, 8. b, 9. a, 10. c, 11. H, 12. F, 13. O, 14. N, 15. D, 16. E, 17. Q, 18. A, 19. G, 20. B, 21. L, 22. I, 23. M, 24. C, 25. K, 26. J, 27. P

16 REPRODUCTIVE SYSTEMS

Chapter Introduction [p.307]

16.1. The Female Reproductive System [pp.308-309]

16.2. The Ovarian Cycle: Oocytes Develop [pp.310-311]
1. ovary, 2. oviduct, 3. uterus, 4. myometrium, 5. endometrium, 6. cervix, 7. vagina, 8. birth canal, 9. labia majora, 10. labia minora, 11. clitoris, 12. urethra, 13. three, 14. menstruation, 15. endometrium, 16. proliferative, 17. endometrium, 18. ovulation, 19. progestational, 20. corpus luteum, 21. endometrium, 22. menarche, 23. menopause, 24. endometriosis, 25. ovarian, 26. meiosis I, 27. ovary, 28. follicle, 29. FSH, 30. zona pellucida, 31. estrogens, 32. secondary, 33. polar, 34. follicle, 35. secondary oocyte, 36. oviduct, 37. fertilization, 38. ovum, 39. endometrium, 40. progesterone, 41. cervix, 42. corpus luteum, 43. endometrium, 44. follicles, 45. implant, 46. endometrium, 47. estrogen, 48. grows, 49. LH, 50. decrease, 51. increase, 52. increase, 53. continue to increase, 54. fully developed, 55. luteal, 56. menstruation, 57. be maintained, 58. corpus luteum, 59. around the middle of

16.3. The Male Reproductive System [pp.312-313]

16.4. How Sperm Form [pp.314-315]

16.5. Sexual Intercourse [p.316]
1. T, 2. before, 3. cooler, 4. T, 5. epididymis, 6. sperm and glandular secretions, 7. T, 8. T, 9. single, 10. e, 11. a, 12. d, 13. c, 14. b, 15. b, 16. c, 17. a, 18. seminiferous tubules, 19. spermatogonia, 20. mitosis, 21. meiosis, 22. primary spermatocytes, 23. meiosis I, 24. secondary spermatocytes, 25. meiosis II, 26. spermatids, 27. sperm, 28. flagellum, 29. Sertoli, 30. head, 31. acrosome, 32. mitochondria, 33. Leydig cells, 34. Leydig cells, 35. Testosterone, 36. testosterone, 37. anterior, 38. hypothalamus, 39. decrease, 40. LH, 41. FSH, 42. increase. 43. fertilization, 44. meiosis, 45. the male's bladder, 46. pancreas, 47. pregnancy, 48. orgasm, 49. at any time in the menstrual cycle.

16.6. Fertilization [p.317]

16.7. Preventing Pregnancy [pp.318-319]

16.8. Options for Coping With Infertility [pp.320-321]
1. F, 2. K, 3. B, 4. D, 5. A, 6. G, 7. N, 8. M, 9. O, 10. I, 11. C, 12. L, 13. J, 14. P, 15. E, 16. H. 17. Aborting a

late-term fetus is controversial unless the mother's life is threatened.; 18. uterus; 19. Capacitation; 20. zona pellucida; 21. meiosis; 22. ovum; 23. fertilization; 24. zygote; 25. One-third; 26. fertility; 27. HMG (human menopause gonadotropin; 28. ovulation; 29. in-vitro fertilization; 30. uterus; 31. intracytoplasmic sperm injection; 32. GIFT (gamete intrafallopian transfer); 33. ZIFT (zygote intrafallopian transfer); 34. 20

16.9. Some Common Sexually Transmitted Diseases [pp.322-323]

16.10. STDs Caused by Viruses and Parasites [pp.324-325]

16.11. Eight Steps to Safer Sex [p.325]

16.12. Cancers of the Breast and Reproductive System [pp.326-327]

1. c, 2. a, 3. c, 4. b, 5. a, 6. b, 7. c, 8. b, 9. b, 10. b, 11. c, 12. a, 13. Pelvic inflammatory disease. May develop after chlamydial infection or gonorrhea. Symptoms include pain and possible sterility., 14. E, 15. B, 16. G, 17. C, 18. D, 19. A, 20. F; 21. breast; 22. BRCA1, BRCA2; 23. mammogram; 24. lumpectomy; 25. radical mastectomy; 26. cervix; 27. endometrium; 28. Ovarian; 29. Testicular; 30. prostate

Self-Quiz

1. d, 2. a, 3. b, 4. a, 5. b, 6. c, 7. c, 8. d, 9. a, 10. b, 11. d, 12. a, 13. d, 14. e, 15. a, 16. b, 17. f

---

17 DEVELOPMENT AND AGING

Chapter Introduction [p. 331]

17.1. Overview of Early Human Development [pp.332-333]

17.2. From Zygote to Implantation [pp.334-335]

17.3. A Baby Times Two [p.335]

1. F, 2. B, 3. C, 4. A, 5. D, 6. E, 7. a. mesoderm, b. ectoderm, c. endoderm, d. mesoderm, e. ectoderm, f. mesoderm, g. ectoderm, h. mesoderm, i. ectoderm, 8. fertilization, 9. zygote, 10. cleavage, 11. morula, 12. gastrulation, 13. cell differentiation, 14. morphogenesis, 15. fold, 16. die, 17. G, 18. H, 19. J, 20. B, 21. F, 22. A, 23. K, 24. D, 25. I, 26. C, 27. E

17.4. How the Early Embryo Develops [pp.336-337]

17.5. Extraembryonic Membranes [p.338]

17.6. The Placenta; A Pipeline for Oxygen, Nutrients and Other Substances [p.339]

1. E, 2. G, 3. D, 4. H, 5. A, 6. I, 7. F, 8. B, 9. C, 10. yolk sac, 11. embryonic disc, 12. amniotic cavity, 13. chorionic cavity, 14. primitive streak, 15. neural tube, 16. future brain, 17. somite, 18. pharyngeal arches, 19. c, 20. b, 21. a, 22. d, 23. ectoderm, 24. tube, 25. but not on, 26. stop, 27. death, 28. f, 29. e, 30. e, 31. d, 32. a, 33. b, 34. a, 35. b, 36. c, 37. f, 38. f, 39. e, 40. f

17.7. The Second Four Weeks: Human Features Appear [pp.340-341]

17.8. Development of the Fetus [pp.342-343]

17.9. Birth and Beyond [pp.344-345]

1. specialize, 2. head, 3. human, 4. gonads, 5. testes, 6. the absence of testosterone, 7. fetus, 8. can, 9. one inch, 10. first, 11. lanugo, 12. vernix caseosa, 13. second, 14. move, 15. third, 16. survive, 17. distress,

18. circulatory, 19. umbilical, 20. placenta, 21. vein, 22. lungs, 23. liver, 24. mother's, 25. collapsed, 26. atrium, 27. foramen ovale, 28. birth, 29. heart, 30. increases, 31. foramen ovale, 32. pulmonary, 33. systemic, 34. close, 35. forebrain, 36. lens, 37. pharyngeal arches, 38. heart, 39. somites, 40. tail, 41. lower limb bud, 42. upper limb bud, 43. umbilical cord, 44. retinal pigment, 45. external ear, 46. foot plate, 47. upper limb, 48. T, 49. smooth, 50. fetus', 51. T, 52. first, 53. T, 54. cervix 55. T, 56. placenta, 57. T, 58. inhale, 59. year, 60. T, 61. colostrum, 62. T, 63. T, 64. uterine

17.10. Disorders: Miscarriage, Stillbirths and Birth Defects [pp.346-347]

17.11. Prenatal Diagnosis: Detecting Birth Defects [p.348]

17.12. From Birth to Adulthood [p.349]

17.13. Time's Toll: Everybody Ages [p.350-351]

1. Improper nutrition can cause birth defects such as spina bifida, as well as affecting a baby's mental development. Taking certain drugs, smoking or using alcohol can also affect the normal development of the baby., 2. C, 3. G, 4. A, 5. F, 6. H, 7. D, 8. E, 9. B, 10. CVS, 11. Amniocentesis, 12. preimplantation diagnosis, 13. Fetoscopy, 14. adolescence, 15. neonate, 16. senescence, 17. infancy, 18. embryo, 19. fetus, 20. zygote, 21. G, 22. D, 23. F, 24. E, 25. I, 26. A, 27. C, 28. J, 29. B, 30. H, 31. Brain tissue contains masses of neurofibrillary tangles and is riddled with amyloid plaques. , 32. The gene for apolipoprotein E. Early-onset Alzheimer's., 33. Aging CNS neurons tend to lose some of their myelin sheath and do not conduct action potentials as efficiently. In addition, neurotransmitters may be released more slowly.

Self-Quiz

1. d, 2. b, 3. b, 4. a., 5. b, 6. d, 7. F, 8. B, 9. J, 10. I, 11. H, 12. L, 13. D, 14. M, 15.A, 16. N, 17. E, 18. C, 19. G, 20. K

---

18 CELL REPRODUCTION

Chapter Introduction [p. 356]

18.1. Reproduction: Continuing the Life Cycle [pp.356-357]

1. J, 2. H, 3. D, 4. C, 5. G, 6. I, 7. B, 8. A, 9. E, 10. F, 11. histones, 12. nucleosome, 13. condenses

18.2. Overview of the Cell Cycle and Cell Division [p.358-359]

18.3. The Four Stages of Mitosis [pp..360-361]

1. metaphase (D), 2. early prophase (C), 3. telophase (F), 4. prophase (E), 5. anaphase (A), 6. transition to metaphase (B), 7.false, 8. false, 9. Prophase, 10. interphase, 11. sister chromatids, 12. centromere, 13. chromosome, 14. condensed, 15. microtubules, 16. centrioles, 17. two pairs, 18. Prophase, 19. chromosomes, 20. poles, 21. chromatids, 22. metaphase, 23. anaphase, 24. chromatid, 25. Telophase, 26. nuclear envelope, 27. number, 28. telophase, 29. mitosis, 30. interphase, 31. mitosis, 32. G1, 33. S, 34. G2, 35. prophase, 36. metaphase, 37. anaphase, 38. telophase, 39. cytoplasmic division,

40.13, 17, 12, 11, 9, 18, 9., 41.C, 42. C, 43. B;
44.diploid, 45. sister, 46. centromere, microtubules,
47. spindle, 48. centrioles, 49. b, 50. a, 51. a, 52. b, 53.
b, 54. b

18.4. How the Cytoplasm Divides [p.362]
18.5. Concerns and Controversies Over Irradiation
[p.363]

1. B, 2. D, 3. A, 4. C, 5. Natural sources include
cosmic rays from outer space and radioactive radon
gas in rocks and soil., 6. Ionizing radiation damages
cells by breaking apart chromosomes and alter]ing
genes. If the damage occurs in germ cells, the
resulting gametes may give rise to infants with genetic
defects. If somatic (body) cells are damaged, the
damage may include burns, miscarriages, eye
cataracts, and cancers of the bone, thyroid, breast,
skin, and lungs., 7. In medicine, X-rays, MRI, and
PET scanning are valuable uses of irradiation.
Irradiation therapy is useful in treating cancer. Food
is irradiated to kill harmful microorganisms and to
prolong shelf life by preventing vegetables like
potatoes from sprouting.

18.6. Meiosis—The Beginnings of Eggs and Sperm
[pp.364-365]
18.7. A Visual Tour of the Stages of Meiosis [pp.366-367]

1. one; 2. two; 3. four; 4. n; 5. gamete formation; 6.
Meiosis I; 7. Meiosis I; 8. Meiosis II; 9. Meiosis II; 10.
Meiosis I; 11. E (2n); 12. D (2n); 13. B (2n); 14. A (n);
15. C (n); 16. 2 (2n); 17. 5 (n) 18. 4 (n); 19. 1 (2n); 20.
3 (n); 21. Spermatogenesis: gamete formation in
males. A single diploid germ cell—the primary
spermatocyte—undergoes meiosis, producing four
haploid spermatids of equal proportions. Oogenesis:
gamete formation in females. A primary oocyte
contains many more cytoplasmic components than a
primary spermatocyte. The primary oocyte undergoes
meiosis to produce four cells of different sizes and
functions. Only one will develop into an ovum; 22.
anaphase II (H); 23. metaphase II (F); 24. metaphase
I (A); 25. prophase II (B); 26. telophase II (C); 27.
telophase I (G); 28. prophase I (E); 29. anaphase I
(D).

**18.8. The Second Stage of Meiosis—New
Combinations of Parents' Traits** [pp.368–369]
18.9. Meiosis and Mitosis Compared [pp.370–371]

1. B; 2. D; 3. E; 4. A; 5. F; 6. C; 7. a. mitosis; b.
meiosis; c. meiosis; d. meiosis; e. mitosis; f. meiosis; 8.
Crossing over during Prophase I produces new
combinations of genes on the same homologous pair.
Random mixing of maternal and paternal
chromosomes occurs at Metaphase I. Chance
determines which sperm fertilizes the egg. 9. A; 10. D;
11. B; 12. F; 13. E; 14. G; 15. C; 16, four; 17. eight; 18.
four; 19. eight; 20. two.

Self-Quiz

1. a; 2. d; 3. a; 4. d; 5. c; 6. e; 7. a; 8. a; 9. d; 10. c; 11. b;
12. c; 13. d.

19 Observable Patterns of Inheritance
Chapter Introduction [p. 375]
19.1. Basic Concepts of Heredity [p.376]
19.2. One Chromosome, One Copy of a Gene [p.377]
19.3. Genetic Tools: Testcrosses and Probability
[pp.378-379]

1. A; 2. C; 3. K; 4. E; 5. I; 6. D; 7. H; 8. J; 9. B; 10.F; 11.
monohybrid; 12. chromosome; 13. segregation; 14.
haploid; 15. Cc; 16. Punnett; 17. probability; 18. don't
have to; 19. doesn't; 20. genotype; 21. recessive; 22. a;
23. b; 24. d; 25. c; 26.a. C and c; b. c; c. C and c go
across the top of the Punnett square; c's go down the
side of the square; d. 1 Cc: 1 cc; e. 1 chin fissure: 1
smooth chin; f. 1/2; the predicted ratio remains
constant for each fertilization event; 27. a. 1/2 x 1 =
1/2; b. 1/2 x 1 = 1/2; c. 1/2 chin fissure and 1/2 smooth
chin; 28. The taster parent is heterozygous, Aa. The
nontaster child must be homozygous and had to have
received one nontasting gene from each of his
parents; 29. The man is cc and the woman is Cc. The
probability that the child will have either a smooth
chin or a chin fissure is 1/2 for each; 30. The woman
of normal pigmentation with an albino mother is
genotype Aa; the woman received her recessive gene
(a) from her mother and her dominant gene (A) from
her father. The woman's husband is aa and will give
each of their offspring an a allele. It is likely that half
of the couple's children will be albinos (aa) and half
will have normal pigmentation but be heterozygous
(Aa)., 31. B, 32. B, 33. A

**19.4. How Genes for Different Traits are Sorted Into
Gametes** [pp.380-381]

1. C; 2. B; 3. A; 4. a. 9/16; b. 3/16; c. 3/16; d. 1/16; 5.a. ¾
; b. ¼ ; c. ¾ ; d. ¼ ; e. ¾ X ¾ = 9/16; f. ¾ X ¼ = 3/16;
g. ¼ X ¾ = 3/16; h. ¼ X ¼ = 1/16.

19.5. Single Genes, Varying Effects [pp.382-383]
19.6. Other Gene Effects and Interactions [pp.384-385]

1. a. pleiotropy; b. multiple alleles; c. codominance; d.
polygenic inheritance; 2. Sickle cell anemia is due to
two copies of the mutant allele – all of the
hemoglobin made is faulty. Sickle cell trait individuals
are heterozygous and produce only half faulty
hemoglobin. Symptoms are much more severe in
those with sickle cell anemia; 3. It can possibly
reactivate genes coding for fetal hemoglobin normally
produced only before birth, thus supplementing the
ability of the red blood cells to carry oxygen
effectively; 4. f; 5. d; 6. f; 7. g; 8. b; 9. a; 10. c; 11. g;
12. e; 13. a.

Self-Quiz

1. d; 2. b; 3. a; 4. d; 5. c; 6. a; 7. a; 8. c; 9. d.

20 CHROMOSOMES AND HUMAN
GENETICS
Chapter Introduction [p. 389]
20.1. A review of Genes and Chromosomes [p.390]

20.2. Picturing Chromosomes with Karyotypes [p.391]

1. J; 2. E; 3. B; 4. F; 5. C; 6. G; 7. I; 8. K; 9. D; 10. M; 11. L; 12. N; 13. A; 14. H; 15. F; 16. A; 17. C; 18. D; 19. B; 20. G; 21. E, 22. C, 23. C, 24. A, 25. A

20.3. The Sex Chromosomes [pp.392-393]

20.4. Human Genetic Analysis [pp.394-395]

1. Top two blocks of Punnett square should be XX and bottom two blocks should be XY; 2. sons; 3. mothers; 4. daughters; 5. nonsexual; 6. D; 7. G; 8. C; 9. F; 10. A; 11. H; 12. B; 13. E; 14. This disorder occurs in females who are heterozygous for the trait. the activated X chromosome with a genetic mutation leads to darker patches of skin while the activated normal X chromosome leads to lighter patches of skin..

20.5. Inheritance of Genes on Autosomes [pp.396-397]

20.6. Inheritance of Genes on the X Chromosome [pp.398-399]

20.7. Personalized Medicine [p.400]

1. b; 2. a; 3. c; 4. a; 5. b; 6. a; 7. c; 8. c; 9. a; 10. c; 11. c; 12. d; 13. d; 14. The woman's father is homozygous recessive, aa; the woman is heterozygous normal, Aa. The albino man, aa, has two heterozygous normal parents, Aa. The two normal children are heterozygous normal, Aa; the albino child is aa; 15. Assuming the father is heterozygous with Huntington disorder and the mother is normal, the chances are 1/2 (50%) that the son will develop the disease; 16. If only male offspring are considered, the probability is 1/2 (50%) that the couple will have a color-blind son; 17. The probability is that half of the sons will have hemophilia; the probability is 0 that a daughter will express hemophilia; the probability is that half of the daughters will be carriers; 18. If the woman marries a normal male, the chance that her son would be color blind is 1/4 (25%); if she marries a color-blind male, the chance that her son would also be color blind is also 1/4(25%); 19.. A mutated gene that would normally code for the production of the protein dystrophin does not function. This protein functions to support muscle fibers, so without it, the muscles break down under physical stresses leading to eventual total muscle destruction., 20. a. autosomal dominant; b. X-linked recessive; c. autosomal dominant; d. X-linked recessive; e. X-linked dominant; f. X-linked recessive; g. autosomal recessive; h. autosomal recessive; i. autosomal dominant; j. autosomal dominant; k. autosomal recessive.

20.8. Changes in a Chromosome or its Genes [pp.400-401]

20.9. Changes in Chromosome Number [pp.402-403]

1. duplication (B); 2. deletion (A); 3. translocation (C); 4. A mutation is a change in one or more of the nucleotides that make up a particular gene. Three possible processes that lead to mutation are: deletion (cri-du-chat), translocation (several rare types of cancer) and duplication (no conditions currently linked to this).; 5. a. aneuploidy; b. polyploidy; c. nondisjunction; d. trisomy; e. monosomy; 6. Most changes in the number of chromosomes (aneuploidy) arise through nondisjunction during gamete formation; 7. c; 8. b; 9. c; 10. d; 11. a; 12. b; 13. d; 14. c; 15. a.

Self-Quiz

1. d; 2. d; 3. b; 4. c; 5. b; 6. a; 7. c; 8. b; 9. b; 10. c, 11. b

21 DNA, GENES, AND BIOTECHNOLOGY

Chapter Introduction [p. 407]

21.1. DNA: A Double Helix [pp.408-409]

21.2. Passing on Genetic Instructions [pp.410-411]

1. Withstand weed killers or make their own pesticides. Give them the same advantages or increase nutritional value 2. A five-carbon sugar (deoxyribose), a phosphate group, and one of the four nitrogen-containing bases found in DNA; 3. a. adenine; b. guanine; c. thymine; d. cytosine; e. double-ring; f. double-ring; g. single-ring; h. single-ring; i. thymine; j. cytosine; k. adenine; l. guanine; 4. four; 5. five-carbon sugar; 6. T; 7. helix; 8. T; 9. hydrogen; 10. T; 11. T; 12. sugar; 13. phosphate group; 14. guanine; 15. thymine; 16. adenine; 17. cytosine; 18. nucleotide; 19. covalent; 20. hydrogen; 21. unit of hereditary information, or sequence of nucleotides coding for a specific polypeptide chain;

22. T - A   T - A
G - C   G - C
A - T   A - T
C - G   C - G
C - G   C - G
C - G   C - G

23. bases; 24. nucleotide; 25. semiconservative; 26. DNA polymerases; 27. repair; 28. two million; 29. mutation; 30. ultraviolet radiation; 31. thymine; 32. xeroderma pigmentosum; 33. gene mutations; 34. base pair substitution; 35. deletion; 36. expansion; 37. inborn errors of metabolism; 38. germ, 39. beneficial, 40. D, 41. B, 42. D, 43. C

21.3. DNA Into RNA:The First Step in Making Proteins [pp.412-413]

21.4. The Genetic Code [pp.414]

21.5. tRNA and rRNA [pp.415]

21.6. The Three Stages of Translation [pp. 416-417]

1. a. deoxyribose; b. ribose; c. adenine, thymine, guanine, cytosine; d. adenine, uracil, guanine, cytosine; 2. transcription; 3. translation; 4. transcription; 5. three; 6. mRNA; 7. translation; 8. a. rRNA: nucleic acid chain that combines with certain proteins to form ribosomes, which are involved in assembly of polypeptide chains; b. mRNA: linear sequence of nucleic acids that deliver protein-building instructions to ribosomes for translation into polypeptide chains; c. tRNA: nucleic acid chain that picks up a specific amino acid and delivers it to the ribosome where it will pair with a specific mRNA code for that particular amino acid; 9. In transcription, only the gene segment serves as the

template—not the whole DNA strand (as in DNA replication). Enzymes called RNA polymerases are involved instead of DNA polymerases. Transcription results in only a single-stranded molecule—not one with two strands (as in DNA replication); 10. E; 11. B; 12. F; 13. A; 14. C; 15. D; 16. AUG UUC UAU UGU AAU AAA GGA UGG CAG UAG; 17. E; 18.C; 19.F; 20.A; 21.D; 22.B; 23. These proteins interact with DNA to speed up or halt transcription of a gene; 24. F; 25. G; 26. H; 27. A; 28. E; 29. D; 30. C; 31. B; 32. a. initiation; b. elongation; c. termination; 33. The tRNA anticodon sequence is UAC AAG AUA ACA UUA UUU CCU ACC GUC AUC; 34. amino acids: met leu tyr cys asn lys gly trp gln stop; 35.A; 36.F; 37.C; 38.E; 39.B, 40. D, 41. A cluster of ribosomes, all translating the same mRNA transcript at the same time; whenever the cell needs many copies of a single protein, 42. Perform functions in the cytoplasm or be modified and shipped to other areas of the cell or outside the cell.

21.7. Tools for Engineering Genes [pp.418-419]
21.8. "Sequencing" DNA [p.420]
1. DNA; 2. bacteria; 3. recombinant DNA; 4. species; 5. replicate; 6. genetic engineering; 7. plasmids; 8. restriction enzyme; 9. "sticky"; 10. clone; 11. amplify; 12. G; 13. F; 14. H; 15. A; 16. D; 17. E; 18. B; 19. C; 20. PCR is used to amplify DNA fragments in a test tube, using primers instead of using bacteria as cloning vectors; 21. These short nucleotide chains base-pair with any complementary DNA sequences. DNA polymerases recognize primers as "start" tags; 22. DNA sequencing; 23. probe; 24. gene library.

21.9. Mapping the Human Genome [pp.420-421]
21.10. Applications of Biotechnology to Human Concerns [p.422-423]
21.11. Engineering Bacteria, Animals, and Plants [p.424]
21.12. To Clone or Not to Clone? [p.425]
1. 3.2; 2. 21,500; 3. 1 ½ ; 4. T; 5. T; 6. T; 7. has; 8. T; 9. T; 10. gene therapy; 11. transformation; 12. transformation, 13. transfection; 14. allele; 15. seven; 16. five; 17. cancer; 18. interleukin; 19. attack; 20. interleukins; 21. suicide tags; 22. gene therapy; 23. DNA fragments; 24. repeats; 25. unique; 26. replace; 27. the genetic engineering of organisms to help recycle or breakdown contaminants such as oil; 28. "Designer plants" will improve agriculture and enhance crop yields; Genetically enhanced "oil-eating" bacteria help with oil spills; 29. transgenic; 30. Plasmids; 31. proteins; 32. insulin; 33. micro-injected; 34. diseases; 35. resistance; 36. Therapeutic cloning; 37. stem cells; 38. human tissues; 39. organs; 40. reproductive cloning; 41. embryo.

Self-Quiz
1. d; 2. d; 3. d; 4. b; 5. c; 6. b; 7. c; 8. a; 9. c; 10. a; 11. d; 12. b; 13. a; 14. b. 15. b; 16. c; 17. a; 18. b

---

22 GENES AND DISEASE: CANCER
Chapter Introduction [p. 429]
22.1. The Characteristics of Cancer [pp.430-431]
22.2. Cancer, A Genetic Disease [pp.432-433]
22.3. Cancer Risk from Environmental Chemicals [p.434]
1. F; 2. G; 3. H; 4. B; 5. E; 6. C; 7. A; 8. D; 9. carcinogenesis; 10. Proto-oncogenes; 11. oncogene; 12. does not; 13. does not; 14. tumor suppressor gene; 15. one; 16. tumor suppressor genes; 17. division; 18. activate; 19. expressed; 20. first mutation; 21. second mutation; 22. uncontrolled proliferation, cancer formation; 23. controlled growth; potential for cancer with a second mutation; 24. controlled growth, no cancer; 25. C; 26. A; 27. D; 28. E; 29. B; 30. F; 31. d; 32. b; 33. a; 34. e; 35. c; 36. b; 37. c; 38. a; 39. e; 40. d; 41. half; 42. 40; 43. pesticides; 44. spraying; 45. industrial; 46. mutations; 47. potential; 48. C; 49. B; 50. A; 51. D; 52. A

22.4. Some Major Types of Cancer [p.435]
22.5. Cancer Screening and Diagnosis [p.436-437]
22.6. Cancer Treatment and Prevention [p.438-439]
1. bowel, bladder; 2. sore; 3. bleeding, bloody; 4. lump; 5. Indigestion; 6. wart, mole; 7. cough, hoarseness; 8. B; 9. F; 10. I; 11. D; 12. G; 13. A; 14. H; 15. E; 16. C; 17. sarcomas; 18. carcinomas; 19. gliomas; 20. Lymphomas; 21. leukemias

Self-Quiz
1. b; 2. d; 3. a; 4. c; 5. b; 6. a; 7. e; 8. b; 9. b; 10. e; 11. b; 12. c; 13. a; 14. d

---

23 PRINCIPLES OF EVOLUTION
Chapter Introduction [p.443]
23.1 A Little Evolutionary History [p.444]
23.2 A Key Evolutionary Idea – Individuals Vary [p.445]
1. Charles Darwin, 2. HMS *Beagle*, 3. Galapagos Islands, 4. Thomas Malthus, 5. Natural Selection, 6. b, 7. e, 8. h, 9. f, 10. a, 11. g, 12. c, 13. d

23.3 Microevolution: How New Species Arise [pp.446-447]
1. There is variation among individuals in a population. These variations can be passed from generation to generation. Some variations of a trait are more advantageous than others. The fittest traits are more likely to be selected for survival ("survival of the fittest"). A population is evolving if variations of traits (alleles) become more or less common. Changes in the gene pool are responsible for biodiversity., 2. species, 3. Mutations, 4. natural selection, 5. *On the Origin of Species*, 6. fundamental principles, 7. adaptation, 8. the founder effect, 9. genetic drift, 10. gene flow, 11. gene flow, 12. fertile offspring, 13. isolating, 14. divergence, 15. gradual speciation, 16. beneficial trait, 17. A & B, 18. B & C, 19. D

23.4. Looking at Fossils and Biogeography [pp.448-449]

23.5. Comparing the Form and Development of Body Parts [pp.450-451]

23.6. Comparing Genetics [p.452]

1. l, 2. d, 3. g, 4. a, 5. f, 6. e, 7. b, 8. h, 9. c, 10. j, 11. i, 12, k, 13. True, 14. True, 15. False, 16. True, 17. False, 18. False

23.7. How Species Come and Go [pp.452-453]

23.8. Evolution from a Human Perspective [pp.454-455]

23.9. Emergence of Early Humans [pp.456-457]

1. d, 2. a, 3. c, 4. b, 5. e, 6. bonobos, 7. primates, 8. prehensile, 9. opposable, 10. forward-directed eyes, 11. trees, 12. omnivorous, 13. bow-shaped; smaller teeth of about the same length, 14. fewer, 15. longer, 16. brain, 17. Culture, 18. language, 19. bipedalism, 20. shorter, 21. bipedalism, 22. anthropoids, 23. climate, 24. great apes, 25. hominins, 26. Lucy, 27. did, 28. *erectus*, 29. 18,000, 30. African emergence, 31. Multiregional, 32. cultural, 33. Eukarya, 34. Animalia, 35. Chordata, 36. Mammalia, 37. Primates, 38. Homininae, 39. *Homo*, 40. *sapiens*

23.10 Earth's History and the Origin of Life [pp.458-459]

1. F, 2. T, 3. F, 4. F, 5. F, 6. T, 7. T, 8. F, 9. T 10. F

Self-Quiz

1. c, 2. a, 3. c, 4. d, 5. d, 6. b, 7. d, 8. c, 9. b, 10. c, 11. b, 12. b, 13. a, 14. b, 15. b, 16. a, 17. d, 18. d

---

24 PRINCIPLES OF ECOLOGY

Chapter Introduction [p.463]

24.1. Some Basic Principles of Ecology [p.464-465]

1. i, 2. e, 3. a, 4. c, 5. g, 6. f, 7. b, 8. h, 9. d, 10. j, 11. a, 12. b, 13. b; 14. a

24.2 Feeding Levels and Food Webs [pp.466-467]

24.3 Energy Flow Through Ecosystems [p.468]

1. producers, 2. autotrophs, 3. consumers, 4. heterotrophs, 5. herbivores, 6. carnivores, 7. omnivores, 8. decomposers, 9. energy, 10. recycled, 11. a, 12. d, 13. e, 14. c, 15. f, 16. a, 17. a, 18. b, 19. top carnivores, 20. carnivores, 21. herbivores, 22. producers, 23. decomposers, 24. input, 25. sun, 26. are, 27. trophic, 28. chain, 29. food webs, 30. plants, 31. lower, 32. producers, consumers, 33. biomass, 34. decreases, 35. primary productivity

24.4 Introduction to Biogeochemical Cycles [p.469]

24.5 The Carbon Cycle [p.470-471]

1. producers, 2. biogeochemical cycles, 3. reservoir, 4. rapidly, 5. slowly, 6. global water, 7. atmospheric, 8. Sedimentary, 9. geologic uplifting, 10. E, 11. F, 12. A, 13. B, 14. C, 15. D, 16. evaporation from ocean, 17. precipitation into ocean, 18. wind-driven water vapor, 19. evaporation from land plants (transpiration), 20. precipitation onto land, 21. surface and groundwater flow, 22. D, 23. G, 24. A, 25. F, 26. C, 27. B, 28. H, 29. E.

24.6 The Nitrogen Cycle [p.472]

1. d, 2. a, 3. c, 4. b, 5. c, 6. b, 7. e, 8. nitrogen gas, 9. nitrogen fixation, 10, nitrification, 11. uptake by autotrophs, 12. food webs on land, 13. denitrification, 14. leaching

Self-Quiz

1. b, 2. b, 3. c, 4. c, 5. b, 6. a, 7. a, 8. d, 9. c, 10. b, 11. d, 12. c, 13.d, 14. d

---

25 HUMAN IMPACTS ON THE BIOSPHERE

Chapter Introduction [p. 476]

25.1 Human Population Growth [pp.476-477]

25.2 Nature's Controls on Population Growth [p.478]

1. billion, 2. Growth rate, 3. birth, death, 4. India, 5. above, 6. developed, 7. sixty, 8. demographics, 9. density, 10. age, 11. base, 12. slow, 13. double, 14. exponential, 15. S-shaped, 16. carrying capacity, 17. high-density, 18. e, 19. f, 20. b, 21. a, 22. d, 23. c, 24. g

25.3 Ecological "Footprints" and Environmental Problems [p.479]

1. population growth, 2. natural resources, 3. ecological footprint, 4. United States, 5. renewable resources, 6. nonrenewable resources, 7. pollution, 8. pollutant, 9. human, 10. point source, 11. nonpoint source, 12. riding a bicycle, 13. using a fan, 14. cutting down evergreens for Christmas trees and replanting with new trees, 15. growing a home vegetable garden

25.4 Assaults on Our Air [pp.480-481]

1. c, 2. d, 3. a, 4. b, 5. d, 6. e, 7. b, 8. b, 9. e 10. c, 11. a, 12. e

25.5 Global Warming and Climate Change [pp.482-483]

1. greenhouse gases, 2. heat energy, 3. greenhouse effect, 4. lower, 5. primary production, 6. photosynthesis, 7. higher, 8. 420,000, 9. global warming, 10. one, 11. global climate change, 12. precipitation, 13. drought, 14. melting, 15. glaciers, 16. sea level, 17. 2100, 18. coastline

25.6 Problems with Water and Waste [pp.484-485]

25.7 Problems with Land Use and Deforestation [pp.486-487]

1.d, 2. h, 3. e, 4. g, 5. b, 6. i, 7. a, 8. j, 9. c, 10.f

25.8 Moving Toward Renewable Energy Sources [pp.488-489]

1. Crossword Puzzle Solution: a. solar power, b. photovoltaic, c. hydrogen fuel, d. coal, e. greenhouse, f. biofuels, g. turbines, h. fossil fuels, i. hybrid, j. ethanol, k. nuclear, l. wind, m. gas, n. core, o. radioactive; 2. petroleum, 3. electric power generation, 4. coal, 5. natural gas, 6. petroleum, 7. Transportation, Residential and Commercial Uses, Industry, Electric Power Generation

25.9 Endangered Species and the Loss of Biodiversity [pp.490-491]

25.10 Biological Magnification [p.491]

1. c, 2. i, 3. b, 4. a, 5. d, 6. f, 7. biological magnification, 8. mosquitoes, 9. body lice, 10. tissues, 11. producers, 12. consumers, 13. predators, 14. top, 15. shells, 16. extinction

Self-Quiz

1. d, 2. a, 3. a, 4. c, 5. b, 6. a, 7. d, 8. a, 9. d, 10. a, 11. c, 12. d, 13. c, 14. a, 15. d